What People Are Saying About

The Science of Free Will

I am delighted to see this work come to fruition. Samir Varma breaks down barriers in this fascinating book on determinism. Even if physics leaves no room for free will in theory, you effectively have "Free Will in Practice" —you are part of the causal chain steering your actions in complex, unpredictable ways. I thoroughly enjoyed this book and believe many readers will gain profound insights from it.

Tyler Cowen, *New York Times* bestselling author of 19 books and Professor of Economics at George Mason University

An intellectual tour de force, this book delves into the timeless debate surrounding free will's existence. Bridging theory and practice, it weaves an exhaustive interdisciplinary study through science, mathematics, computer science, and philosophy. With compelling evidence and insightful caveats, it challenges readers to reconsider the boundaries between determinism and human agency.

Silas Beane, Professor of Physics at the University of Washington, renowned for his work in theoretical nuclear physics and computational science

A thought-provoking and stimulating read on fate, destiny, and free will, with fascinating examples that will keep you engaged and yearning to learn more! Physicist, entrepreneur, and hedge-fund manager, Samir Varma, draws upon physics, computation, and plain logic to present some surprising and persuasive connections!

Jay Kesan, PhD, JD, distinguished professor, highly cited legal scholar, accomplished author, and a res
with extensive expertise in intellectua

The Science of Free Will is a wonderfully entertaining case for determinism. I learned a lot about wave functions, computability and Artificial Intelligence. Was I convinced that I lack free will? Well, no. But that can hardly be blamed on the author! Convinced or not, I learned a lot, enjoyed the book, and I predict you will too!

Alex Tabarrok, Chair in Economics at the Mercatus Center, George Mason University, and noted researcher in political economy

Dr. Varma brilliantly connects physics and philosophy as they always should have been. The breadth of knowledge covered in this book can only be done by a true intellectual.

Dan Amiram, Dean at Coller School of Management, Tel Aviv University, and noted business advisor

A cognitive joyride! Physicist and philosopher Samir Varma takes readers on a whirlwind adventure prowling through quantum mechanics, neuroscience, and artificial intelligence to reveal why free will persists even in a deterministic universe.

Stephen J. Salzer, award-winning ENT Surgeon and "Top Doctor" per Castle Connolly

A thrilling intellectual adventure exploring one of life's greatest mysteries—do we really have free will? This book takes readers on a rollercoaster ride through cutting-edge physics, avant-garde computer science, and mind-bending philosophy to resolve this age-old paradox once and for all.

Salman Khan, Founder and CEO, Stabilis Capital Management, LP

What does determinism have to do with driving in traffic? Can bees get PTSD? Does the Universe offer choices? Physicist and entrepreneur Samir Varma deftly explores these and other surprising connections between free will, computation, and the fundamental laws of physics. Whether out of choice or as destiny's destination—this book is a must-read.

Priya Sahgal, author, talk show host, and Editorial Director, NewsX

What does physics reveal about fate and destiny? Does chaos theory spell the end of free choice? This electrifying book explores new frontiers where quantum mechanics and AI crash headlong into philosophy's oldest question.

Rasheed Kidwai, noted author, journalist, and Visiting Fellow at the Observer Research Foundation

Think you have free will? Think again. In this provocative page-turner, physicist and entrepreneur Samir Varma probes deep into the laws of physics, the nature of computation, and cutting-edge neuroscience to reveal how our fates may not be our own—yet why we still have freedom in ways that matter for a meaningful life.

Suresh Mandava, author, award-winning surgeon and ophthalmologist

Samir sharpens our understanding of free will to prepare us as we step into a machine intelligent world.

Jitendra Kavathekar, Technologist, Executive, Investor

Guilt. Accountability. Hard work. Success. Romance. Corporate strategy. War. Ethics. Individual purpose. These and many other critical facets of human existence are essentially meaningless without free will. But if all actions in the universe are dictated by unquestionably deterministic physical laws beginning at the subatomic level, how can nondeterminism, choice, and unpredictability arise? In this seminal tour de force, Samir Varma takes us on a rich, thought-provoking journey spanning quantum physics, biology, mathematics, theology, law, computing, economics, and philosophy to help us gain deep insights into our universe—and ourselves.

Joe Weinman, bestselling author of *Cloudonomics* and *Digital Disciplines* (an Amazon #1 Hot New Release), inventor, and IT executive

The Science of Free Will

How Determinism Affects Everything
from the Future of AI to Traffic to
God to Bees

The Science of
Free Will

How Determinism Affects Everything
from the Future of AI to Traffic to
God to Bees

Foreword by Tyler Cowen, *New York Times*
bestselling author of *The Great Stagnation,*
The Complacent Class, Average Is Over, and *Big Business*

Samir Varma

IFF
BOOKS

London, UK
Washington, DC, USA

CollectiveInk

First published by iff Books, 2025
iff Books is an imprint of Collective Ink Ltd.,
Unit 11, Shepperton House, 89 Shepperton Road, London, N1 3DF
office@collectiveinkbooks.com
www.collectiveinkbooks.com
www.iff-books.com

For distributor details and how to order please visit the 'Ordering' section on our website.

Text copyright: Samir Varma 2023

ISBN: 978 1 80341 731 8
978 1 80341 743 1 (ebook)
Library of Congress Control Number: 2023951800

A CIP catalogue record for this book is available from the British Library.

Design: Lapiz Digital Services

UK: Printed and bound by CPI Group (UK) Ltd, Croydon, CR0 4YY
Printed in North America by CPI GPS partners

We operate a distinctive and ethical publishing philosophy in
all areas of our business, from our global network of authors to
production and worldwide distribution.

Contents

List of Figures with Captions

To Chase, my beloved four-legged boy who left us much too soon, and to Edmund Sullivan, Lisa Fulton, and Lindsay Thalheim, the three vets that pushed the boundaries of technology to try to save him. He taught us what being good really means.

Preface

The most fundamental paradigm we have for understanding ourselves, and the world we live in, is deterministic: the laws of physics. Physics is, by definition, the most fundamental set of statements we can make about the universe as a whole, which necessarily includes ourselves as we too are part of the universe. This book traces the logical consequences of that determinism with a particular focus on the concept of free will which has been a bone of contention for millennia. Additionally, many things are more simply stated, without loss of generality, using computing paradigms rather than physical laws. Therefore, I discuss determinism in both contexts: physics and computing.

By taking the determinism of the laws of physics seriously, we can talk about an amazing range of things. Do bees get PTSD? What does MacOS tell us about the workings of the human brain? What do Isaac Asimov's three laws of robotics tell us about Artificial Intelligence? Can AI be conscious? What is science? Why is traffic so annoying? Why don't even Nobel Prize winning economists understand taxation? What is a minimal computer and just how minimal can it be? What type of God is compatible with modern science? What are morality and ethics in a deterministic world? What does the future of AI have to do with the market for used cars? Can cockroaches have free will? And much more!

Determinism also has consequences. In chapter 5, I use those consequences to make predictions about the future of determinism, the future of free will, and the future of AI. The future of AI, in particular, is going to lead to some very serious ethical conundrums. Those ethical conundrums are not just theoretical; they will have significant business and societal implications.

I had great fun writing this book, and it is in that spirit that I present it to you. I hope that some of my joy and fun in writing it has seeped through into the text!

S.V.

Foreword

I am delighted to see this book come to fruition.

Determinism is a very important issue that we do not like to discuss, so I am happy to see Samir Varma breaking down these barriers. His arguments are interesting and rigorous throughout.

I have been corresponding with Samir Varma for many years now, stemming from his initial outreach. He has been a reader of my blog, Marginal Revolution, and our connection started when he began to send me links. His links were uniformly excellent in their quality, and so I kept on responding to Samir. I even read the links on physics that he sent along, although I rarely understood them.

Eventually Samir and I met for lunch a few times in northern Virginia. Since the beginning, our correspondence has become extensive. Since our first correspondence on July 9, 2009, he has e-mailed me more than 10,000 times, and I have e-mailed him back more than 7000 times.

Samir Varma is not a professional author or academic, but he has produced a work that is more interesting than most of what you see from those in the academy, or from purely popular writers. I believe that is because Samir wrote this book for himself, and for the potential readers that he respects most. That said, the final product is extremely readable and to the point. It does not bog down in esotericism. Instead it communicates potential truths about our world, truths of great importance.

Is there room in the cosmos for free will at the theoretical level? Samir will tell you, "No." I suspect he may be right. Physics just doesn't leave room for that kind of indeterminacy. Nonetheless we have what Samir calls "Free Will in Practice," namely that you are part of the causal chain that steers your actions, and furthermore the details of what you will choose are virtually impossible to predict. For all practical purposes, you are computationally irreducible. If that is not enough free will for you, well very likely that is all you are going to get. Live with it, as indeed you can.

So I guess I had to write this foreword, but did you know that? One thing I do know is that many of you will enjoy this book very much and learn a great deal from it. Beyond that, I am going to keep the rest of my likely false predictions to myself.

Tyler Cowen

Acknowledgments

I had thought of the solution to the free will conundrum presented in this book some time ago, and always meant to write it up as a book. But I procrastinated: there was always something else to do. At some point, I mentioned casually in an e-mail to Tyler Cowen (https://www.marginalrevolution.com) that I knew how to, once and for all, end the "free will" debate—how can we have free will if we are subject to physical laws and yet those physical laws are deterministic? Tyler insisted that I had to write it up as a book. I resisted. He twisted my arm. I resisted some more. He twisted harder. My general proposition is that when a really smart guy tells you that you **must** do something, you do it and don't ask too many foolish questions. I therefore yielded and I started writing it! But even then, it was going slowly—I suffer from migraines and those are heavily exacerbated by electronic screens: phones, laptops, TVs, etc. Luckily, I got introduced to a truly outstanding neurologist, Jen Werely, that managed, somehow, to control the migraines using a brand-new class of medications called CGRP inhibitors. I had no side effects at all—the migraines just stopped. It was amazing. This meant I had all sorts of time back that I had never had the last 50 years. So then I started seriously thinking about actually finishing this book but was still slightly procrastinating. On the way back from watching a Paul McCartney concert, I mentioned to another friend of mine, Steve Salzer (a brilliant and pioneering ENT surgeon: https://www.greenwichent.com/our-practice-and-doctors), that I was thinking of finishing this book and he urged me to do it! In reasonably certain terms! So now I had a second really smart guy telling me I had to write this thing. Steve also volunteered to be my first editor, sounding board, and proofreader. So, well, I started writing it up! I want to thank both Steve and Tyler for their help and encouragement. Steve's ideas and inexhaustible energy have massively improved this book. And I want to thank Jennifer Werely, the neurologist that figured out what to do about my migraines! Without them, this book would never have been written. I also want to thank Ben Southwood (https://www.worksinprogress.co/issue-author/ben-southwood/) and Jason Crawford (https://rootsofprogress.org) for extremely useful feedback and comments. In particular, both of them suggested I address the issue of compatibilism (see section *1.4, Compatibilism*) and its relationship to what I am proposing. Needless to say, all errors and infelicitous prose are still my responsibility!

Finally, I thank my family for their love and support. Even though it was inevitable that I would end up with them, I am still pleased to have my family: of course my pleasure is also inevitable!

Introduction

Jane's fist slammed down on the table, rattling her coffee mug. "I don't care what physics says!" she yelled in the crowded cafe. "I know I have free will."

Her outburst drew sidelong glances from nearby patrons. Jane hardly noticed, stewing over the implications of determinism she'd learned in her quantum physics class. According to her textbook, every particle making up her body followed exact physical laws, their motions predetermined. This meant in theory, all Jane's choices were illusions—the universe unfolding on rigid rails. She may feel like picking coffee over tea, but that decision was set in stone at the Big Bang. Jane's mind reeled. She hadn't chosen to major in physics only to be told her entire notion of free will was a farce! Gripping her mug tightly, Jane made a decision. She would prove the physicists wrong by demonstrating her ability to make choices. When the barista called her name for her scone order, Jane marched up and blurted out, "I changed my mind. I want a chocolate chip cookie instead of a blueberry scone."

As she bit into the warm, gooey cookie, Jane smiled with satisfaction. Let's see physics calculate the taste of that choice, she thought. Her outburst may have seemed random to the coffee shop, but it made perfect sense to Jane.

On her drive home, Jane cranked up the radio while contemplating the physics lesson that had bothered her so much. Her professor had calmly explained that complete knowledge of every particle in the universe would allow predicting Jane's actions with absolute certainty. To her professor, Jane slamming her fist on the table was just the inevitable result of subatomic particles bouncing around. Jane grinned, remembering the bewildered look on the barista's face when she changed her scone order to a cookie order on a whim. But her smile faded as she wondered—could even a spontaneous choice be predetermined? Over the next few days, Jane designed a kind of experiment to test free will. She jotted down five random tasks on slips of paper: baking cookies, reading her astrophysics textbook, going for a run, painting her nails, calling her brother. After shuffling them, Jane would close her eyes and draw one slip. Whatever task she picked, she would have to complete, no matter how she felt about it. Closing her eyes, Jane drew "call your brother." She heaved a sigh. She and her brother had fought recently and hadn't spoken in weeks. The thought of calling made Jane's stomach turn. Jane hesitated, gripping the slip of paper tightly. Surely she could draw again or pick a different task instead? But she had designed this experiment to force her into an unpleasant choice she would not make otherwise. Steeling herself, Jane dialed her brother's number. Physics may try to deny her free will, but she would prove she had it.

Jane's brother picked up on the third ring, his voice hesitant. "Hey, sis, everything okay?" Jane fidgeted with the paper slip in her hand. "Yeah, I just … wanted to talk," she stammered.

They started with stilted small talk about the weather and work. Then Jane took a deep breath. "Listen, I'm sorry about our fight. It was dumb to argue over politics like that.

You're my brother and that's what matters most."

Her brother was quiet for a moment. "I'm sorry too," he said finally. "I hate fighting with you. Truce?"

Jane smiled, feeling tension uncoil in her chest. "Truce."

After they hung up, Jane tidied up the slips for her experiment. She had been tempted to abandon the unpleasant phone call, but instead saw it through based on her own resolve. Surely the ability to overcome preferences and make difficult choices defied the clockwork determinism of particles. In her next physics lecture, Jane raised her hand to pose a question. "Isn't the unpredictability of human actions proof that we have free will?" The professor paused to consider this. "Some believe so," she said carefully. "But there could be a level of complexity to particles' interactions that we cannot calculate, even if the interactions themselves follow exact laws." Jane furrowed her brow. The professor's response echoed her own lingering uncertainty. Perhaps free will was possible in practice, even if physics denied it in theory. After her conversation with the professor, Jane decided to visit her computer science friend Neha to learn more.

"Have you heard of computational irreducibility?" asked Neha. She explained, "Even simple sets of mathematical rules can produce unpredictably complex results. The only way to know what will happen is to run through every step—you can't shortcut the computation."

Jane's eyes lit up with understanding. "So even if physics says determinism is true in theory, the complexity makes outcomes unpredictable in practice—like free will emerging from particles!"

Neha nodded. "Exactly. Some call it 'apparent free will.' The universe is deterministic, but irreducible complexity means we can never calculate everything ahead of time."

Jane leaned back in her chair, feeling vindicated by this concept. Perhaps the mechanics of her choices were predetermined on some subatomic level. But as far as her daily life went, nothing could predict her whims—whether she would have cookies or salad for lunch, or whether she would call her brother when angry or make amends. She could never anticipate her own actions either. That interplay of determinism and surprise was free will enough for Jane. Just because physics rendered the future unpredictable didn't make the present any less meaningful.

"Thanks, Neha," Jane said with a smile. "I think I've got my answer." She now saw free will not as some mystical power, but as the natural result of irreducible complexity at work in the world.

What Jane realized is that the combination of two well known facts leads to a big problem. Fact #1: Physics is deterministic. Fact #2: We humans are made up of atoms, and those atoms are following deterministic physical laws. Neither fact is anything other than banal. But put them together, and you have a giant conundrum: what is free will?

Chapter 1

What Is Free Will?

The problem is this: we are made up of atoms. Our bodies are primarily made up of Carbon, Hydrogen, Oxygen, Nitrogen, Calcium and Phosphorus. But these are the exact same atoms as the rest of the universe! They're the same atoms that make up planets, and galaxies, and the sun, and rest of the universe. Every one of those atoms is made up of still more fundamental entities, namely electrons, protons and neutrons. And protons and neutrons are made up of still more fundamental entities called quarks and gluons. So, at bottom our bodies are made up of electrons, quarks, and gluons. But here's the thing: every one of those fundamental particles is obeying precise mathematical laws, which we have discovered over the last 100 years. So, if every single entity that makes up our body is following precise mathematical laws, then in what sense do we have free will? The net result of this is a large, sprawling, and unsettled argument where neither side, neither the anti-free will side nor the pro-free will side, convinces the other. In this book, I aim to do three things. First, I will explain that both sides are right, but they are both talking about different things—so they're talking past each other. Second, I will explain another cleaner, simpler, and better paradigm—computation—that finally resolves these issues and does so permanently. And third, once I have cleared up the confusion and explained the computational paradigm, I will apply what we have learned and make some predictions.

I will start by bridging the gap between the two warring parties: those who say that free will exists, and those who say it does not. I will argue that the parties are, in some sense, talking about different things. Those who do not believe in free will are talking about "Free Will In Theory" (FWIT), and those who do are talking about "Free Will in Practice" (FWIP). I believe both parties are correct. We both have free will, and simultaneously, we do not. Is this a paradox? The answer is no. In reality, we *do not* have free will in theory but we *do* have free will in practice. In fact, I believe and will argue, something a little bit more extreme: we have exactly no FWIT but have perfect FWIP. In effect, we can and should hold both views simultaneously because they apply to different levels of explanation. To the extent that you can predict someone's behavior *exactly and always* that person cannot, by definition, have free will either in theory or in practice. Therefore, for someone to have free will, their behavior must be, to some extent, unpredictable. To the extent that you cannot predict someone's behavior (including being unable to *predict your own behavior*) that person has free will. The more predictable someone is, the less free will they have. At the point at which they are completely predictable they have no free will at all. As I will show in the chapters that follow, the net result of thinking this way is that even if you have no free will in theory, you have and will forever have, free will in practice. This is conceptually very similar to the standard theory of how much information is contained in a signal. One result of this theory is that signals with more information are less predictable: signals with highest information content are the least compressible because compressibility depends upon predictability. See section *2.12, Signal Compression* for a simple example.

The organization of the book is as follows.

I will first sketch out why free will cannot exist in theory. While doing so I will explain why some of the usual criticisms (for example, the collapse of the wave function is nondeterministic and so that is the origin of free will) are not good arguments. I will then explain why free will does, and in fact *must*, exist in practice. Then I will talk about the practical consequences of FWIT being untrue. And I will argue that there are no practical consequences to FWIP being true since that is the status quo.

I'll also talk about some related things, for example the principle of computational equivalence (chapter 3), that are not strictly necessary for the truth or falsity of my argument, but are illuminating nonetheless because they give additional reasons and motivations to believe what I believe.

Also, this isn't a book on sociology or anthropology or philosophy although it necessarily touches on those topics, which means that I am not going to bother to engage with previous authors in these and other related fields. I'm interested in laying out what I think is a watertight argument backed up by as much evidence as I can gather and am completely uninterested in arguing about someone else's opinion whether it be St. Augustine or Jacques Derrida.[1] Ultimately, you will accept my argument ... or you won't. The only way to find out is to read on until you reach the predetermined (but unpredictable) conclusion!

1.1 The Argument

My argument is simple at its core. Every single thing in the universe is made up of particles which are excited states of quantum fields (section *1.3, Fields*). That is exactly, precisely, literally, everything. Rocks, trees, bedbugs, cockroaches, stars, humans, water, hydrogen, etc. *Everything* is made up of excited states of quantum fields and nothing more. Every single thing you see around you or that exists is made up of particles, each of which has a "wave function". The wave function for any "composite object" such as an atom or molecule is the composite of the wave functions of each of its constituent parts.[2] The theories that explain the workings of these particles, i.e. quantum field theory and quantum mechanics—together referred to as quantum theory—are entirely, completely, and totally deterministic. The solution to a given quantum theory is its wave function which tells us everything that we are ever able to know about the given system. The Schrödinger equation from quantum theory, which determines how quantum states evolve in time, is *entirely* deterministic. This means that the evolution of wave functions (which are the solutions of the Schrödinger equation) is entirely deterministic. The onus of proof therefore is on those asserting that FWIT exists, not on those that say that it doesn't. Why? For one very simple reason. For FWIT to really exist, *something* must be nondeterministic. There are only two logical possibilities for those who claim that FWIT exists. One, those claiming that FWIT exists need to show something, *anything*, in the fundamental laws of physics that is both nondeterministic and can serve as a source of free will in theory. Or, two, they need to assert on faith with no proof of any kind whatsoever, that the fundamental laws of physics are fundamentally wrong or incomplete and contain an undiscovered element of indeterminacy.[3] Obviously, neither of these arguments is rigorous in any way. The first argument is inconsistent with all known science, and the second relies on faith and contradicts known science.

An excellent example of anti-FWIT thinking comes from the eminent physicist Brian Greene:

> "We are made of these exquisitely ordered, wonderfully choreographed particles of nature governed fully by the physical laws, no free will whatsoever," Greene said. "And yet even in that environment, our particular arrangements can through a flitting burst of activity, create beauty, illuminate mystery, experience wonder."
> —Brian Greene, quoted in [77]

The counterargument to this is beautifully explained by Philip Ball:

> But is free will really undermined by the determinism of physical law? I think such arguments are not even wrong;

they are simply misconceived. They don't recognize how cause and effect work, and by attempting to claim too much jurisdiction for fundamental physics they are not really scientific but metaphysical.... If the claim that we never truly make choices is correct, then psychology, sociology and all studies of human behaviour are verging on pseudoscience. Efforts to understand our conduct would be null and void because the real reasons lie in the Big Bang.... The underlying problem here is that the reducibility of phenomena — which is surely valid and well supported — is taken to imply a reducibility of cause. But that does not follow at all. What "caused" the existence of chimpanzees? If we truly believe causes are reducible, we must ultimately say: conditions in the Big Bang. But it's not just that a "cause" worthy of the name would be hard to discern there; it is fundamentally absent. [13]

My argument is that both arguments are completely correct and both are incomplete. Free will really, truly, doesn't exist in theory since it is "running on" the laws of physics: i.e., a perfectly deterministic substrate.[4] Just like you cannot get an "ought" from an "is", you cannot get nondeterminism from any possible combination of deterministic physical laws. But, free will does really exist in practice, *for every possible and conceivable practical purpose*. Why? Because it is *impossible* to calculate anything about a large collection of quantum fields (such as the collection makes up a person). And furthermore, even if it were possible to calculate everything about the large numbers of interacting quantum fields that make up a person, we still couldn't predict anything because to make a prediction about what someone is going to do requires not just perfect information about their internal state, but also perfect information about every single interaction they have with their environment: every molecule of air, every photon of light, and not just at one instant, but potentially also for some number of instants beforehand. This also answers the objections of those who are worried about the effect of what they call "physicalism" on society:

> Somehow, we know we are free to act. It goes beyond intuition. It is basic. Yet, the physicalists say it is an illusion. [...] [T]he implications of this on society are immense. If we are not responsible for our actions, what is justice? Why would anyone reasonably be accountable for their actions if what they did was set in stone by the laws of physics that mysteriously came into play at the time of the Big Bang? [14]

Therefore, *free will is the absence of being able to predict what will occur* (which is what I am calling FWIP).

There is a second, independent, reason as well for this unpredictability. Stephen Wolfram has put forth the concept of Computational Irreducibility: the inability to shortcut a computer program to find out its result, without running the entire program for however long it takes. [143] He has shown that even extremely simple programs produce results that are computationally irreducible—that is, they cannot be shortcutted. You must run the programs to find out what they are going to do (see section *3.15, Irreducibility versus Chaos*). We must go through all the necessary steps to cause whatever was going to happen to happen, and we cannot know in advance what will happen. Therefore, while FWIT is impossible because we are running on a deterministic substrate, FWIP is not only possible but *required* because of Computational Irreducibility: we cannot find out what someone (or even ourselves) is going to do until that person has done it. Furthermore, to know what I will do in the future requires me to simulate both myself, *and* the entire light cone[5] of events around me, while still maintaining my separation from the simulation.[6] This is an impossible task by definition.

> *Your free will is your (as well as everyone else's) inability to predict what you are going to do until you do it!*

1.2 Humans

What is a human? In total, approximately $6.71 * 10^{27}$ atoms in an adult male weighing 70kg. A human is made up of mostly water (i.e., 2 Hydrogen atoms and 1 Oxygen atom) plus a bunch of "organic molecules" which consist of Carbon, Hydrogen and Oxygen combined with other elements. 99% of the body is made up of only six elements: "Hydrogen (62.9%), Oxygen (almost 24%), Carbon (nearly 12%), Nitrogen (nearly 0.6%), Calcium (0.24%) and Phosphorus (0.14%)" [125] and a smattering of other elements. The crucial point is that it's just standard stuff. It's the same standard stuff that makes up the Earth (not surprising since we evolved from earth stuff), but also Mars, Jupiter, the Sun, our galaxy, other galaxies, the stuff of intergalactic space—in fact, everything in the known universe. Many of the remaining trace elements in our bodies were formed in the interior of stars and then dispersed when those stars blew up, i.e., became supernovas. The supernova itself creates still more elements: all the elements heavier than iron are produced during the supernova event. We are all (to use Carl Sagan's famous phrase) "star stuff." Now, since it is all standard stuff, that standard stuff is following the laws of physics. Or, if you think about it "computationally", humans are a program running on a substrate that obeys the laws of physics: the substrate is the individual atoms that make up a human. All atoms[7] of Oxygen are exactly identical to all other atoms of Oxygen, and the same is true for Hydrogen, Carbon, Helium, Strontium, etc. *It is the arrangement of those atoms*—the arrangement itself and only the arrangement—*that identifies a human as a human.* Different humans have different arrangements of their atoms: that is, fundamentally, what makes them different to each other. Different animals have different arrangements of atoms than humans, as do rocks, trees, insects, birds, stars, galaxies, and your favorite Twinkie bar. But they're all made of the same "star stuff!"

All those atoms are made up of particles which we have named electrons, protons and neutrons,[8] and they are held together by force carrying particles called "bosons". So, coming down by levels, humans are made of molecules, which are combinations of atoms. Atoms are made of particles, all madly interacting with each other and with themselves by exchanging bosons. Now what is a particle?

1.3 Fields

To understand what a particle is, we have to first understand "fields". Pervading the entire universe are things called fields which are of many different kinds. A field is an entity that pervades all of space and time (together conventionally called "spacetime"), and to each point in spacetime it associates a set of numbers. Physicists have discovered rules for how those sets of numbers behave. Depending on the type and quantity of numbers you associate with each point in spacetime, you get a different field. Each of the known fields is associated with a specific type and quantity of numbers at each point in spacetime. The set of rules for a specific type of field is its "field theory". A subset of all field theories comprises "quantum field theories" (QFTs) and there is a QFT associated with each kind of particle. A particle is nothing more than an "excited" state of some quantum field described by the appropriate QFT. (An excited state means that the quantum field is not in its vacuum state but in some state that is *not* the vacuum. Physicists call the lowest possible energy state of a quantum field its "vacuum".

Because of Heisenberg's uncertainty principle, quantum fields always have a lowest energy which can never be zero.) It is important to understand that a particle is not "described" by an excited state of the appropriate QFT: it *is* the excited state of the appropriate QFT. So, at the most fundamental level, everything in the universe is made up of Quantum Fields. Every single molecule in the human body is made up of atoms which are made up of fundamental particles (all of which are identical—you cannot tell one electron apart from another) and those fundamental particles are all manifestations of physical laws: the laws of quantum field theory. Finally, here is the kicker:

Quantum Field Theory is entirely deterministic!

One could object and argue that fields describe reality but are not the entities that make up reality. This would be a good counter argument, except that we already know that all of nature is explained by fields. The fields and reality are, at "worst", in one-to-one correspondence: they're what mathematicians call isomorphic.[9] Reality may be made up of them or they may just describe reality. It does not matter. Since we know that fields properly describe reality, reality itself is deterministic since the fields are deterministic—if reality wasn't deterministic, then deterministic fields could not describe it. We can actually go a little further: suppose you have a universe made up of fields, and another universe where the fields are isomorphic to reality but that universe is, in some sense, not made up of fields. And let us assume that both universes have the same physical laws. Given that, how would you design an experiment to distinguish between the universe where reality is made up of fields and the other in which reality is isomorphic to fields? It is clearly not possible. Therefore, since you cannot distinguish between the two, they are actually the same thing. In fact, if there was an experiment to distinguish these two cases, the two cases would no longer be the same: we would no longer be speaking of an isomorphism but an approximation.

Another objection is that many predictions in physics are probabilistic. So where does the probability come from? Isn't that a source of free will? The short answer is that there are *two* basic theories in theoretical physics from which all other physics descends: quantum field theory and general relativity. general relativity is explicitly deterministic: no probabilities there. What about quantum field theory(QFT)? QFT is entirely deterministic too with the *one* and *only one* proviso of a phenomenon that is usually called "wave function collapse" (which is what supposedly happens, in the "Copenhagen interpretation". when we make a measurement—only one of many possibilities is actually detected). Wave Function collapse as a supposed phenomenon is, phrasing it as politely as I can, absurd.

That collapse, *and only that collapse*, provides probabilities in fundamental physics. However, apart from being absurd, that collapse is entirely out of the control of any person or experimenter and *cannot* be the source of free will. I address the absurdity of the "collapse" as well as its inability to be the source of free will in much more detail in section *2.14, Wave Function Collapse*. Every other "probability" (or randomness) in physics comes from us averaging over effects that are either too difficult to calculate from first principles or where we make explicit approximations to allow us to do calculations.

1.4 Compatibilism

There is a section of the literature, called "compatibilism", that seeks to ground free will in a deterministic universe. Various objections have been made to compatibilism called "incompatibilism", which can be stated as follows: [96]

- If someone acts of her own free will, then she could have done otherwise.
- If determinism is true, no one can do otherwise than one actually does.
- Therefore, if determinism is true, no one acts of her own free will.

In response, compatibilists have made various arguments, generally along the lines of: "it doesn't matter if the substrate is deterministic, the person that emerges from the substrate is still responsible for their own actions." One argument is that free will is doing what one wants to do and not doing what one doesn't want to do. If what one wants to do or not do arises from deterministic physical laws, so what? It doesn't mean that you did something you didn't want to do or didn't do something you did want to do. To this, incompatibilists argued that you cannot change history: those facts that have happened have happened. You cannot change the laws of nature: they are what they are. No one can change future facts (i.e., future history) and all such future facts will flow from the laws of nature. Therefore, no one has any control over future history, and therefore there is no free will. In response, compatibilists argued that just because a person wasn't free to choose a different course of action (in my language, they were unable to make another choice because FWIT doesn't exist), it doesn't mean that they were not responsible for their actions. To which incompatibilists raised other objections. To which the compatibilists replied. And so on and on. That debate has continued and shows no signs of abating. The argument in this book shows that the debate, which is perfectly valid, is unnecessary, *if you take literally what modern physics is telling you*. Both compatibilists and incompatibilists should be happy (I predict, of my own free will in practice, that they won't!) with the formulation in this book because, in effect, compatibilists are talking about FWIP, and incompatibilists are talking about FWIT. They're *both* right because they are both talking about different things!

1.5 Accuracy

If we are to believe the formulation I am proposing, quantum field theory better be right! So how accurate is quantum field theory? There are essentially 3 quantum field theories: one for the electromagnetic interaction, called quantum electrodynamics (QED), a QFT for the weak interaction, and a QFT for the strong interaction called quantum chromodynamics (QCD). In the 70s, by using the Higgs mechanism, Steven Weinberg joined together QED and the QFT for the weak interaction. So at much higher energies than we live at, there are actually only two QFTs: the electroweak QFT and QCD. The standard model of particle physics consists of all 3 of those theories and it explains all the interactions between all known particles. While it doesn't explain dark matter particles (although, *assuming dark matter exists*, there is no doubt that to explain dark matter, we will end up with another QFT), that isn't relevant for our purposes because biology and chemistry are unrelated to dark matter. For our purposes, the part of quantum field theory that is most relevant is QED: the quantum theory of light and ordinary matter. It is important to understand and absorb the following point: *all of chemistry* and therefore *all of biology* comes from the interaction of electrons, protons, neutrons, and photons. In other words, all of the human and biological world is explained by QED. QED is probably the most accurate scientific theory ever devised by man.

To give you an example, let's talk about the "anomalous magnetic moment" of the electron. The magnetic moment of an object is a measure of how strong the magnetic source is. So the higher the magnetic moment, the stronger the source. The effectively "classical" result,[10] also called (for technical reasons) the "tree level" result, is calculated via the Dirac equation and is exactly 2. At tree level, we are simply calculating the result without considering "quantum corrections" to the value. We can then use QED to calculate "corrections" to this factor coming from the fact that the electron is interacting with itself in a myriad of ways. As we consider more and more ways in which the electron interacts with itself, we get smaller and smaller corrections to this value. These corrections are called the "anomalous" magnetic moment. So if we write the magnetic moment as g then what we are calculating is how different g is from 2, and we call that difference (divided by 2) the anomalous magnetic moment, a. In other words, we calculate $a = (g - 2)/2$ and then compare that to what we measure when we do an experiment. Now, here is one of the very biggest triumphs of science ever. Our theoretical calculation of a agrees with experiment to one part in 1 billion! And to do that, we had to measure g to one part in 1 trillion! Our calculations show that $a = 0.001,159,652,181,643(764)$ whereas experiment shows that the value is $a = 0.001,159,652,180,73(28)$ (in parentheses is the uncertainty in the value). [142] This is the most accurately verified prediction in the history of physics. But, folks, that's not all! If you look at the Wikipedia page, *Precision Tests of QED*, you will see test after test of QED's predictions in multiple ways using different methods with which to probe its predictions. [152] One of the very coolest tests is that of the so-called "fine structure constant", usually called alpha, that measures the strength of the electromagnetic interaction. Alpha is expected to be constant throughout the universe. Michael T. Murphy and his collaborators [104] measured the spectra of 17 nearby stars, up to 160 light years away, with properties very similar to the sun. They have set an upper limit of just 50 parts per billion on how much alpha may vary between the stars! In other words, alpha is constant precisely as expected. QED is probably the most stringently tested and accurate physics theory of all time.

1.6 Fundamentalism

Yes, that title is tongue in cheek. But it contains a kernel of truth. Physicists think differently than essentially everyone else, including other scientists. The question in physics is always: what is the most minimal explanation possible, what is the precise range of phenomena to which this explanation applies, and how precisely can we predict? So, for example, an electron at rest always has a mass of 9.109×10^{-31} kilograms. No electron at rest can ever have a mass that is *different* than that. Similarly, a photon will *always be massless* and will *always* move at the speed of light when in a vacuum. There are *no exceptions*. The theory of quantum electrodynamics is the theory of light and electricity. That theory will *always* be true.[11] No other scientific field (except perhaps some kinds of chemistry) works this way. That is because physics is *fundamental*. It underlies *everything*. There are *no* exceptions. I am belaboring this point because, while it is second nature and entirely obvious to physicists, it is neither second nature nor obvious to anyone else. In every other field, you can play the "but what about" game. You can't do that in physics. It is really the case that every electron that has ever existed is identical; that every photon that has ever existed is identical; and that the theory that describes them is the *entire* explanation. *There are no exceptions.* It's important to understand that, if you claim that there is something that physics doesn't explain, what you are really saying is that the laws of physics are *incorrect* or *incomplete*. And yes, we know areas of physics where the laws are incomplete. The laws are incomplete at very high energies far outside human experience, which is why we need humongous machines like the LHC (large hadron collider) to complete them. They are also incomplete at the edge of black holes where gravity becomes very strong. But, other than that, *they're complete!* All of biology—let me please repeat that—*all of biology is completely contained within already known laws of physics*. Those laws are *never* going to change. As new physics is discovered, *unlike every other field of human knowledge*, the new laws *must* reduce to the already known laws at these energy scales. I emphasize again: *they must!* For example, in the realm in which they are true, Newton's laws are *still valid!* Physicists use them all the time. What happens is that as we discover new "things", we understand the domain of validity of the old "things". So, for example, when Einstein discovered special relativity, the physics world realized that the domain of validity of Newton's Laws was when speeds were much less than the speed of light. Next, Einstein discovered general relativity, which meant that Newton's laws were only valid when the gravitational field was weak. But, in cases where the gravitational field is weak and the speed is slow, which covers *all of human existence* and almost all of the universe, Newton's laws are entirely valid! When you take the "limit" of Einstein's laws, i.e., when gravity is weak and the speed is much less than that of light, you recover Newton's laws. *You must recover Newton's Laws in that limit else Einstein's Laws would be wrong!* The entire point of this book is to ask what physics can tell us about free will and related issues precisely *because* all the physics underlying biology is *already known*. And, let me beat this dead horse some more: if you say that there is "something" else (i.e., you deny physicalism) you are, *by definition*, saying that the laws of physics are incorrect or incomplete *at the biological energy scale*. You are (at least in the free world) entirely entitled to your own opinion about everything. But if you take that stance, please realize that you are taking a stance that is entirely at odds with well known, totally non-controversial, well settled and *incredibly well tested* physics. Oh, and just by the way, it's also against biochemists who are generally of the view that there is nothing in special in living matter that isn't mere chemistry—to believe otherwise is to believe in the highly discredited idea of "vitalism". So, if you wish to deny physicalism, you're not just far out on a limb, you are so far out that you're on a tiny twig far out at the end of a subbranch of a subbranch of a subbranch of that limb. If that's where you wish to live, that's up to you!

1.7 Physicalism Denial

If, contrary to all evidence, you deny physicalism, you can end up in all sorts of logical tangles. And not just theoretical ones, but practical ones with real consequences. One of the very worst results of not understanding physicalism is a decision by the US Supreme Court: the so-called "Alice" decision. To understand the sheer scientific illiteracy of this decision, we need to first understand a little bit about US patent law. Section 101 of the US Code is called "Inventions patentable" and says:

> Whoever invents or discovers any new and useful process, machine, manufacture, or composition of matter, or any new and useful improvement thereof, may obtain a patent therefor, subject to the conditions and requirements of this title. [2]

This is a deliberately broad and very reasonable definition of what is, and what isn't, patentable. It clearly understands that at the root of everything, there is physics. Machine and manufacture are obviously physical. So too is "composition of matter". But what is a process? The most general definition of a process is that it is a series of actions directed to some end and/ or a continuous action, operation, or series of changes taking place in a definite manner. [45] In other words, a process tells you *how* to do something: it is an algorithm or recipe. That is a physical thing because even a mere thought is a movement of electrons and chemicals in your brain. Until something physical changes in your brain, no thought has actually occurred. Similarly, an algorithm or process has to be instantiated in something physical—as words that you vocalize that turn into sound waves which compress and expand air as they travel to the ears of listeners from which electrical signals flow to their brains, or as electrons storing charges in some computer memory, or as chemical inks on a physical page, and so on. So for there to be a thought, process or algorithm, there *must be* some physical "container" (for lack of a better word) in which it exists—it's "instantiation". In other words, section 101 says, "If you can think it up, you can patent it"! Now, this is subject to some qualifications. Section 102 [3] says that the invention must be novel, and section 103 [4] says that the invention must be non-obvious. Those are straightforward to understand—they mean in law what they mean in general to us. Section 112 [5] requires that the invention must also be enabled, which means that "someone skilled in the art to which it pertains" can make and use it. These are all completely reasonable, and it is easy to understand why a patent must satisfy these conditions. After all, if you produce something so obvious that someone else could have easily thought of it, or you produce something that isn't novel, then there isn't any reason for the government to grant you a monopoly right over that. And if your instructions are so deficient that someone couldn't follow them and make the invention, then what have you really invented? This was more or less the state of affairs since the founding of the republic because Article I Section 8 Clause 8 of the US constitution says:

> Patent and Copyright Clause of the Constitution. [The Congress shall have power] "To promote the progress of science and useful arts, by securing for limited times to authors and inventors the exclusive right to their respective writings and discoveries." [137]

In 2014, however, the scientific-knowledge-free justices of the US Supreme Court stepped into this with a decision so breathtakingly stupid (for lack of a better technical term) that it almost defies description. In Alice Corp. v. CLS Bank

International they held that Alice's patent was invalid because it was an "abstract idea".

> In any event, we need not labor to delimit the precise contours of the "abstract ideas" category in this case. It is enough to recognize that there is no meaningful distinction between the concept of risk hedging in Bilski and the concept of intermediated settlement at issue here. Both are squarely within the realm of "abstract ideas" as we have used that term.[12]

Many people that were opposed to software patents celebrated this decision. Because, after this decision, a very large number of software defined inventions were no longer patentable.

Others lamented the lack of clear guidance as to what is, and what isn't, patentable. Business and legal questions are not really my concern. However, the utter idiocy of the decision is based on a lack of understanding of basic science. It is akin to an English Ph.D. not understanding that Shakespeare was a playwright or an engineer not knowing how to do basic algebra. *Every single thing written in a patent is an abstract idea*: it is *mere information* that is supposed to be novel, non-obvious and enable someone else to build the claimed invention. That information is embodied in physical form as ink on a page, or as pixels on a computer screen, or as an algorithm in computer code, or as patterns of neuronal firings in someone's brain! It is *impossible* to distinguish an "abstract idea" from any other "idea" because their physical forms are identical. The correct standard is exactly what is laid down in sections 101, 102, 103 and 112: that it be "any new and useful process, machine, manufacture, or composition of matter" and it be novel, non-obvious, and enabled. These criteria make sense — they distinguish one class of ideas from another class of ideas. But the truly insane "abstract idea" concept is no different than the "I know it when

I see it" standard.[13] It is meaningless and scientifically illiterate to the point of it being literal nonsense.

The real fault here is that the Supreme Court didn't understand that, at bottom, everything is physical. There is no "platonic" thought out there somewhere. To conceive of something, anything, there has to be something that is doing the conceiving! It is a denial of physicalism. The strange thing is that all of this is so entirely obvious to physicists that no one even thinks or talks about it—it's in the air physicists breathe, and the water they drink. But, under the influence of pre-scientific and nonscientific philosophy, it seems entirely possible to the utterly science illiterate that there can be a thought without a thinker or an idea without an ideator. It is telling that not a single Supreme Court judge has had any scientific training—not even as an undergraduate. It is exactly as if they shied away from learning anything about one of mankind's greatest and hardest won achievements: real, testable knowledge from science. It is after all much easier to have fact free opinions rather than do the very hard work of learning what the facts are and what they mean!

Acres of ink have now been spilled trying to interpret this decision. No one knows what it means, and the Supreme Court, having caused a needless disaster, has consistently refused to take any more cases to clarify what it meant. A cynic[14] would say that this was a deliberate decision, designed to make sure that large firm interests are further entrenched and to make sure to further disadvantage the "little guy". And wouldn't you know it, out comes a paper that exactly explains the effects of the Alice decision!

We examine the impact of lost intellectual property protection on innovation, competition, acquisitions, lawsuits and employment agreements. We consider firms whose ability to protect intellectual property (IP) using patents is weakened following the Alice Corp. vs. CLS Bank International Supreme Court decision. This decision has impacted patents in multiple areas including business methods, software, and bioinformatics. We use state-of-the-art machine learning techniques to identify firms' existing patent portfolios' potential exposure to the Alice decision. While all affected firms decrease patenting post-Alice, we find an unequal impact of decreased patent protection. Large affected firms benefit as their sales and market valuations increase, and their exposure to lawsuits decreases. They also acquire fewer firms post-Alice. *Small affected firms lose as they face increased competition, product-market encroachment, and lower profits and valuations.* [emphasis added] They increase R&D and have their employees sign more nondisclosure agreements. [6]

But it all really boils down to a lack of understanding of science, and a complete lack of intellectual humility and honesty to admit that mistake and correct it. The final irony is that the patent in question was almost certainly invalid under any combination of sections 102, 103 and 112. It was entirely obvious, not novel in any way, and didn't enable "someone skilled in the art to which it pertains" to build or practice the claimed invention. In other words, this ruling was unnecessary—the court could simply have invalidated the patent—but lacking both scientific knowledge and the intellectual humility to know what they didn't know, they produced an absolute clunker of an opinion.

Notes

1. Alex Tabarrok points out, correctly, while commenting on a paper by Hanno Sauer [123], that far too much time is wasted in philosophy discussing the views of old philosophers that could not know all the development of knowledge

that took place after their time. Since those views are not informed by, for example, scientific developments that followed their time, their views are of little value. [131]

2. See section 2.13, *Unitarity – Unitary Evolution.*

3. To add to the spiciness, quantum mechanics is the only theory in science that has never had a single observation, ever, that is contrary to its predictions.

4. I discuss determinism in chapter 2.

5. A light cone is simply the furthest distance from which a photon could reach an observer in a given time period — it represents the maximum possible extent in space that is able to affect the observer. Anything further away is said to be "outside the observer's light cone" and it therefore cannot affect the observer.

6. Because the simulation needs to run independently of me.

7. More precisely, all isotopes of Oxygen. An isotope is an atom with a different number of neutrons in its nucleus. Oxygen has 18 isotopes: ^{11}O to ^{28}O. The superscript is the "atomic number" which is the sum of the number of protons and neutrons in the nucleus. Oxygen is *defined* by having 8 protons in its nucleus. Therefore, ^{11}O has 3 neutrons, $11 - 8 = 3$, up to ^{28}O which has $28 - 8 = 20$. Of these only ^{16}O, ^{17}O, and ^{18}O are "stable": that is, they do not decay. Almost all the Oxygen in the universe is ^{16}O. What I'm saying is that every single isotope of Oxygen is *exactly* the same as every other isotope of Oxygen that has the same number of neutrons.

8. Protons and neutrons are made up of still more fundamental particles called quarks and gluons.

9. Wikipedia has a nice discussion of isomorphism: see reference [148].

10. Classical means non-quantum-mechanical.

11. At the energies at which it is applicable, which easily encompasses all biological phenomena.

12. 134 S.Ct. at 2357 (citing Bilski v. Kappos, 561 U.S. 593, 611 (2010)).

13. "In his concurring opinion in the 1964 Jacobellis v. Ohio case, Supreme Court Justice Potter Stewart delivered what has become the most well-known line related to the detection of 'hard-core' pornography: the infamous 'I know it when I see it.' statement." See [93].

14. i.e., me!

Chapter 2

Determinism

Determinism means that given a complete physical state at time T_1, you can *exactly* determine the physical state at time T_2 via physical laws. To do this you must know everything about the physical state at T_1. In the case of trying to predict the future state of a human, to do this would mean knowing the exact configuration of every subatomic particle that makes up every atom that makes up every molecule that makes up the human at time T_1. It *also* means knowing the configuration of every subatomic particle that may impinge upon that specific human during the time interval from T_1 to T_2. Since photons are the fastest things that can exist, and they move at the speed of light, it means that for every second that ticks by, the configuration of every subatomic particle within one light-second of that person must be known. For reference, light travels at about 3×10^8 meters/second, which means that one light second is a distance of 3×10^8 meters, or 3×10^5 kilometers = 3 hundred thousand kilometers![1]

Therefore, in principle, if we wanted to predict what a human might do one second from now, then if we knew about every subatomic particle's configuration within a 3 hundred thousand kilometer radius of that human, we could make a prediction. Within 3 hundred thousand kilometers of any human is all of the earth (every single human, ocean, blade of grass, animal, insect, chemical, factory, etc.) plus a whole bunch of space. The moon, for example, is approximately 380 thousand kilometers away on average, so we are talking about a sphere with a radius that almost encompasses the moon's orbit! And this does not even consider that a computer capable of doing this prediction would be of greater complexity and size than the system it is simulating, and any conceivable computers doing this calculation would take longer to do this calculation than would the universe, also called waiting and letting what is going to happen, happen![2] Let us call this the "Practical Limit to Prediction" argument.

Even if we knew all this, we would run into a situation where the results are computationally irreducible—in other words, the only way to find out what the system would do is to just run the system since no "shortcut" to the final result exists.[3] This is related to the problem of why people keep pets. Biologically speaking, a pet is a mechanism to release Oxytocin in the brain which is good for our health, promotes love and bonding, and makes us happy.[4] However, the interaction with a pet seems to be computationally irreducible. Because, if it wasn't, we could simply simulate interacting with a pet in our own brains and get all the benefits with none of the downsides (vet expenses, dead dogs, cats stuck in trees, etc.). But we can't simulate interacting with a dog inside our brains, not just because we do not have the computing power, but also because the interaction itself is computationally irreducible. There is no "shortcut" calculation that our brain can do that will enable us to simulate the interaction with a dog within it. Let's call this the "Computational Irreducibility Limit to Prediction" argument. By the way, this is a two-way street. As far as your dog is concerned, you are a mechanism for releasing Oxytocin in its brain.

"We found that dogs shed tears associated with positive emotions," says Takefumi Kikusui of Azabu University in Japan. "We also made the discovery of oxytocin as a possible mechanism underlying it." [118]

This does not rule out approximate predictions, because to the extent that predictions can be made, they will necessarily be approximate. As discussed previously, if we *could* predict exactly, then free will is ruled out *both in practice and in theory*. However, the ability to predict *approximately* does not rule out free will in practice, as free will remains precisely the portion of the outcome that cannot be predicted: the "error" in the prediction. And when it comes to individual humans, this "error" is huge, as we all know from our own experience: *all* of that is our free will in practice. This leads to a very interesting issue, pointed out to me by Ben Southwood. Suppose we compare two people: one normal (whatever normal means for a human) and one with some kind of syndrome that affects their behavior. Let's call them "N" and "C", respectively. Frequently, the predictability of C will be different than N: C could be more predictable, or less predictable. So the question is: is the free will (in practice) of C different from the free will of N? The answer to this is very interesting and very subtle. First, we are all fundamentally unpredictable and will *always* remain so. So from that point of view the free will of C is no different than N. On the other hand, if we allow for approximate predictability, then yes, in some sense, C has a very very very tiny bit more (or less) free will than N depending on if C is less (or more) predictable than N. The point is that C's free will comes from the collective behavior of all the atoms that make up C's body interacting with C's environment in the same way that N's free will comes from the collective behavior of all the atoms in N's body interacting with N's environment. If, somehow, the atoms that make up C's body are *approximately* more predictable than N's, then, yes, in some sense, their collection of atoms has a very tiny bit less free will than N. Why very tiny? Because when we're talking about 10^{27} atoms in an average human body, the difference is really an immeasurably minuscule amount.

One interesting point to consider is this: how do we know that reality is deterministic? The answer is that we know this from experiments. No matter which experiment we use to measure the mass of an electron, we always find it to be $9.1093837 \times 10^{-31}$ kilograms, plus or minus some measurement uncertainty. Similarly, no matter how we do the measurement, we find that photons always travel at the speed of light, 299,792,458 meters/second, again plus or minus some measurement uncertainty.[5] No experiment or observation exists that contradicts these numbers. Similarly, whenever we hit a tennis ball in the air, it always falls to the ground. And (neglecting air resistance) it always falls to the ground while accelerating at the same rate: 9.80665 (meters/second) per second. The moon still revolves around the earth according to Newton's laws. Light passing close to the sun is deflected according to Einstein's general relativity. In fact, all of science rests upon the repeatability of experiments. *That repeatability of experiments (and in the case of astronomy, repeatable observation) is what tells us that reality is deterministic.*

2.1 Intuition

Determinism means that FWIT is impossible. One reason for finding that hard to accept is our intuition. But one must be careful with using intuition. Our intuition, almost always, comes from equilibrium considerations and from linear thinking. We find systems that are out of equilibrium or nonlinear hard to think about. Interestingly, life itself is a nonequilibrium system.[6] It has to be to extract energy from its environment and use that for its own purposes. And yet, strangely, we create nonlinearities for ourselves all the time. Nonlinearity means that a very small change can produce a very large effect. In sports, for example, we create lots of nonlinearities because we find them exciting. A close game may have no statistically significant difference between the winner and the loser. Yet, we demand a winner and a loser. Think about tennis: you could have a score like 0-6, 0-6, 6-4, 6-4, 6-4. The loser will have won 24 games, (6 + 6 + 4 + 4 + 4), and the winner will have won only 18, 0 + 0 + 6 + 6 + 6!

And, since points you win in games that you lose don't count, you can (and do) frequently have a winner of a match who lost more points than the winner! We want some points in tennis to be more important than others because we find that exciting.

Our intuition generally consists of extrapolating whatever the previous trend was: this stock has gone up, so it must continue to go up; this entrepreneur was successful last time, so he is likely to be successful this time, etc. And when we think further, we tend to do the opposite of the simple extrapolation: a gambler says that his luck is about to turn, or, the market has gone down so much, it has to go up now. We evolved to avoid being eaten by lions, in which case a linear extrapolation of recent trends (the lion is chasing me, so I better keep running for it will keep chasing me), or a simple opposite of the recent trend (the lion has been chasing me for some time so I will soon be able to stop because I am a better long distance runner than any other animal) will work very well. We didn't evolve to make careful estimations of rational risk and reward scenarios.

The same applies to out of equilibrium systems. Our thinking tends to rely on equilibrium considerations, when all the factors that go into consideration are reasonably balanced against each other. For example, journalists love having "balanced" points of view to give you "both sides of a story". Yet, they will do this even when one side of a story is utterly absurd and/ or is scientifically impossible—we simply prefer equilibrium scenarios. Nonlinearity also has one further effect: it makes predictions extremely difficult. The US Electoral College for electing presidents is a great example of this since the popular vote winner doesn't necessarily win the election. And yet, almost all polling is done to estimate the popular vote because it is too difficult, time consuming, and error prone to keep polling individual states. On the night of the US presidential elections of 2016 and 2020, the electoral futures markets swung wildly, trying to discount the outcome of the election. The election probabilities for Clinton v Trump and Biden v Trump were swinging from an almost certain win for one to an almost certain win for the other and back again as the markets tried to price in probabilities as election results came in. If the election were simply about the popular vote, this swing would never have happened—there is almost no nonlinearity in the overall popular vote, but there is plenty of nonlinearity when the election is spread over 50 states, where a single vote can swing a state. Who is going to forget the bizarre images of "hanging chads" in Florida, when trying to decide the 2000 presidential election? Election officials got into almost existential arguments about how much of a hole in the ballot paper counted as a vote! That was a perfect illustration of a nonlinearity: whoever took Florida, even if by one vote, won the election. And so, as one might expect, that which we find counterintuitive tends to be highly nonlinear. Even when you know what is happening, nonlinearity can be incredibly frustrating, irritating, and upsetting. My favorite example of this is my own experience with traffic. I truly detest traffic, driving in traffic, waiting in traffic, everything about traffic. But the kind of traffic that makes me most annoyed is the one that doesn't have an obvious reason: there is no crash, lane closure, obstacle, weather, or any obvious reason. In this case, the cause of the traffic is perfectly well known. What happens is that once the volume of traffic exceeds a certain threshold, any slight perturbation such as a driver hitting the brakes accidentally, changing lanes slightly carelessly, or driving at even a slightly inconsistent speed, will cause a slowdown wave: one car slows down, so the next slows down, then the next and so on. Human reaction delays then make the wave bigger. Pretty soon, everyone is slowing down way behind where one driver did something mildly out of the ordinary. This behavior is surprisingly easy to model and understand (for instance, see reference [56]). And yet, despite understanding the mathematics perfectly well, I find it absolutely infuriating. And, via introspection, it's fairly clear that it is the nonlinearity of the behavior—the fact that some small perturbation caused the large incident—that is so frustrating. Our brains seem to have evolved to want large results to flow from large changes, and small results to follow from small changes. Unfortunately for our brains, the world turns out not to be so! It is quite interesting to look at a traffic simulation. A very simple example of a traffic simulation is the "Rule 184" Cellular Automaton. I discuss cellular automata in chapter 3, *Computation*, and section *3.18, Traffic! Traffic! Traffic!* has an illustration of modeling traffic using Rule 184.

2.2 Nonintuitive Nonlinearity

This section's title is probably an oxymoron. Given how humans think, just about all nonlinearities are nonintuitive or counterintuitive. My favorite example of the nonintuitiveness of nonlinearity is from economics: the "Laffer" curve. It has caused acres of controversy, and even Nobel laureates in economics have not understood it. And yet, as I will show below, the Laffer curve is a tautology: it comes out of incredibly simple mathematics. Given how much controversy it causes, I will rather belabor the explanation below yet, please be assured, this doesn't require even basic arithmetic to understand. If you know how to read a graph, and know how to multiply by zero, you will immediately see that there cannot possibly be any controversy about it, and yet there is! The Laffer curve is the relationship between a *tax rate* and the *total amount of money raised by that tax*. It's just a sequence of points: at tax rate t_1 you will raise $\$d_1$, at tax rate t_2 you will raise $\$d_2$ and so on. We are going to be physicists and, like Albert Einstein, conduct a "gedankenexperiment": a thought experiment. For concreteness, let's think about the income tax. Suppose that the income tax rate was zero percent. How much revenue would the government raise from the income tax? That's entirely obvious:

zero! If there is no tax, there is no revenue raised by it. Now, *suppose that the income tax rate was 100%!* Yes, I mean that if you earn $1, the government takes the *entirety* of that $1. How much revenue would the government raise? Anyone? Anyone? The answer, not quite so obvious, is *zero!* Why? Would you work for *no money at all? None. Nada. Zippo?* Remember, this is a thought experiment and we are specifying no loopholes, no set asides, no deductions, nothing.[7] So, we now know, with perfect certainty, two points on this "Laffer" curve. If we draw the curve so that the income tax rate is on the horizontal (x) axis, and the revenue raised at that tax rate is on the vertical (y) axis, we can plot these two points. We put a dot on the bottom left of the graph: 0% income tax produces $0 revenue. We put a dot on the bottom *right* of the graph: 100% income tax produces $0 revenue *as well*.

So, belaboring this because it is controversial, but shouldn't be, we have two plots of the curve plotted. Bottom left, and bottom right. Now, *by definition*, government revenue, at any other tax rate, has to be at or above zero. So the curve rises from the bottom left, up and to the right, in some unspecified way. And the curve returns, in some unspecified way, back to zero at the bottom right. This we know, and I repeat, we know with *100% certainty. It mathematically has to be this way.* You are not allowed an opinion on this—it is a fact, and it is a fact of simple arithmetic. This curve, since it must rise and then fall back down, is *nonlinear*: it is *not* a straight line. It bends back on itself, and it can bend back on itself any number of times. Now, try this on a piece of paper to convince yourself. Go on, get a piece of paper and a pencil—I'll wait. OK, good, you're back? Draw two axes and plot our two fixed points. Now, draw *any* curve you like that joins our two fixed points, and is always above the x axis. I repeat, draw *any* curve that joins those two points and remains above the x axis (because government revenue cannot fall below zero). Now, you may have drawn a simple curve with one peak, or another curve with multiple peaks. Or something truly hideous and squiggly. It doesn't matter. Pick any point now on the y axis representing the amount of government revenue you wish to raise. Draw a line horizontal to the x axis through that point and note where that line crosses the curve that *you drew!* You will note that *unless you drew the horizontal line at the exact peak of the curve you drew, the horizontal line crosses your curve at least twice!* In other words, you have just proved that for a given amount of government revenue, there are always *at least two tax rates, one high and one low,* that will raise the same revenue. *So it is emphatically not true, contrary to what you see constantly argued, that cutting tax rates must lead to a loss of revenue by the government.* Whether cutting taxes raises or lowers government revenue depends upon the shape of this curve, and where you are on this curve when you cut or raise taxes.[8] For reasons that I used to find mystifying, this is controversial. But I no longer find the controversy mystifying—it is a straightforward consequence of nonlinearity (because the curve bends back on itself) intersecting with ideological and motivated reasoning.

There is still further insight to be gained from this thought experiment. That has to do with *whether or not maximizing the government's tax take* is a good idea in the first place. Everyone acts like it is. But another, extremely simple, gedankenexperiment will convince you that it's far more complicated, and needs nuanced thought. Remember that if you tax something more, you get less of it. This is true by definition, since the tax raises the price, and so the amount demanded must fall. So that means that money raised for the government to spend has, as a direct cost, a lessening of economic activity, and a lessening of economic growth. This *must* be true. On the other hand, you cannot have economic growth in an anarchy: you need laws and regulations, they must be sensible (i.e., not captured by interest groups seeking rents), and they must be enforced impartially. That is the role of a government, and that government needs to have tax revenues to spend to enforce them.[9] This *also must* be true! So we have two intercompeting forces: one that increases the size of the economy, and the other that decreases it, but *both need each other to exist!* And so, yet again, we know the following. Suppose that the tax rate was zero. How would the economy as a whole fare? Well, the answer is, very badly. There would be no revenue for a government to spend, and so there would be no government. So, as a first approximation (and a very good one), economic growth would be very close to zero. Similarly, try a tax rate of 100%. And, yes, you guessed it, economic growth would be horrible as there would be (barring cheating) no reason to work. So as another first approximation (and another good one), economic growth would be very close to zero. And now we do the exact same thing as last time: draw whatever curve you like that connects the zero on the bottom left, to the zero on the bottom right. Go ahead, I'll wait again. I'm very patient. This curve bends back on itself as well, just like the previous one. It too has a "peak" (or peaks) that represents the rate that maximizes economic growth. Now, here is the point. As a general rule in mathematics, there is no relationship between the optima of two different, independent, mathematical functions. In other words, there is *no reason whatsoever that the tax rate that maximizes revenue is the same tax rate that maximizes economic growth!*

There are effectively three nonlinearities here, which is why this seems so nonintuitive. There is the nonlinear curve that shows you the relationship between government tax revenue and tax rates. And there is a nonlinear curve that shows you the relationship between government tax revenue and economic growth. Those are *different* curves. The third nonlinearity is to realize that growth compounds: it is "compound interest" not simple interest. In other words, no matter how small you start something initially, if you wait long enough for it to compound, it will eventually grow very large when compared to how it would have grown were the interest not "compounding". So the *long run* tax rate that maximizes government revenue *will always be* the tax rate that maximizes *economic growth* not the tax rate that maximizes *short term* government tax revenue.[10] These are simply mathematical facts—there is nothing to argue about and there should be no controversy of any kind about them. But the fact is that they are all nonlinear, and that because nonlinearity interacts with our morals, cultural inclinations, and partisanship mean that a controversy is generated where none should *ever* exist!

This doesn't even account for that fact that government is not perfect. Unless you maintain that the way the government spends tax money is *the most efficient way in which it can spend on its citizens' welfare*, you have a fourth nonlinearity to account for: the tax money raised by the government has a cost to the economy in that the money is being spent by the government in ways that are not *optimal*. To the extent that government spends money inefficiently, those are resources that are being wasted. This is a *necessary and inescapable* consequence of any non-optimality in government spending.[11] So the next time someone calls the "Laffer Curve" "voodoo economics" (to use George H.W. Bush's famous phrase) please realize that he was talking through his hat and so is everyone else that doesn't understand (or perhaps want to understand?) nonlinearity. The correct questions you should be asking are always: i) what tax structure will maximize economic growth in the *long run* (and you can define growth in whatever way you want—it doesn't matter—use GDP, don't use GDP, adjust GDP for environmental effects, effects on wildlife, climate, whatever you want);[12] and ii) what is the best way for society to spend those resulting tax revenues? Those are the real questions; the questions that matter. That is what we should be arguing about in a civilized society. Everything else is noise, partisanship, and a basic lack of understanding of nonlinearity.

2.3 Economic Growth

As soon as you internalize that the laws of physics apply to everything, you also realize that economic growth is, effectively, unlimited. Why? Because all we have in our lives is the atoms that we can access. For now, we can only access those atoms that are close to the surface of the Earth, but soon we should be able to access those that are in the deep oceans and in space. Even computation, which people tend to think of as something abstract, consists of the movement of physical objects—currently we use electrons but we could use almost anything—photons, molecules, atoms, etc. (US Supreme Court, are you listening? You are scarily clueless about science. See section *1.7, Physicalism Denial.*) So what that means is that *all economic growth consists of the rearrangement of physical objects—atoms, molecules, photons, electrons, and so on in ever more useful ways.* There are an absolutely incredible number of atoms just here on Earth: 10^{50}. The number of arrangements of 10^{50} atoms is, very roughly, $10^{50}!$ where ! means "factorial". A factorial just means multiply together all numbers up to that number. So 5! means $1 \times 2 \times 3 \times 4 \times 5 = 120$, 6! means 720, 7! is 5040, 20! is 2432902008176640000, and 25! is 15511210043330985984000000. Factorials grow incredibly quickly! $10^{50}!$ is such a large number that we have no method by which to represent it! It is so large that it's beyond conventional comprehension and can't be represented easily. It would have an enormous number of digits and isn't a value that's typically worked with or has practical relevance in most contexts. We might as well think of it as infinite. And that means that we can never run out of growth. And that is why growth pessimism is false! It is, for all intents and purposes, impossible to run out of new ways to rearrange the atoms that we can access!

2.4 Morality and Ethics

One result of thinking with FWIP is realizing that it allows us to shed many fears and superstitions. There can only be two sources of morality and ethics: hardcoded in the brain and/or resulting from the interplay of humans with each other and the broader environment. We can see that this must be true: after all, we are made of the precise same stuff as the rest of the universe. It manifests itself as a pattern of molecules that we call human, and morality and ethics are simply specific patterns of the firing of neurons in our brains, which we can imperfectly translate to communicate with each other. Once we see this, we realize that there is nothing amazing or odd about elephants mourning their dead, or dolphins trying to help drowning humans by nudging them to keep them afloat. They too are made up of the same stuff as the rest of the universe, and they too are exhibiting signs of their own morals and ethics which also result from a combination of their hardcoding and their interplay with their environment.

A growing number of behavioral studies, combined with anecdotal observations in the wild—such as an orca pushing her dead calf around for weeks—are revealing that many species have much more in common with humans than previously thought. Elephants grieve. Dolphins play for the fun of it. Cuttlefish have distinct personalities. Ravens seem to respond to the emotional states of other ravens. Many primates form strong friendships. In some species, such as elephants and orcas, the elders share knowledge gained from experience with the younger ones. Several others, including rats, are capable of acts of empathy and kindness. [18]

2.5 Gödel's Incompleteness

Gödel's incompleteness theorem says, roughly, that in any axiomatic system there will always be true theorems that we cannot prove to be true. (An axiomatic system is a set of specified rules—the axioms—and from those we derive new statements that logically follow, called "theorems".)

Gödel's theorem can be thought of as saying, *there is a state of the system that can be reached by applying its rules but there is no way to reach that state by some shortcut that avoids us having to go through the entire process of applying the rules, and that process could involve an **infinite series** of steps.* A prediction is, by definition, a "shortcut" to a result achieved by not running the entire system to see what the result would be. If you predict that it will snow tomorrow, you are using some set of über-rules (theorems) that shortcut running the entire system (also known as waiting until tomorrow) to find the result. Gödel's incompleteness theorem just means that we cannot always predict in practice using the rules of the axiomatic system.

This does *not* save free will in theory. Determinism is simply the statement that there is a set of rules that is always followed. It is not the statement that those rules need to have certain properties or *that those rules must allow you to prove every true theorem implied by those rules.* All determinism is saying is that some set of rules must be followed and that all outcomes are the result of following those rules and nothing else.

If the argument is that a deterministic system can be modeled as a set of mathematical axioms to which Gödel applies, then all Gödel is saying is that a deterministic system can produce undecidable propositions. And so if we have to take this seriously as the source of free will, we are stating that it is the existence of undecidable propositions that produces human free will. And, as we have already seen above, this does not follow.

One interesting issue related to Gödel is about the usefulness of infinity. Why do we need proofs? Why do mathematicians spend so much time and effort on constructing them? Yes, they're beautiful in their own right, and yes, creating them is fun, and very absorbing work. But why, really, do we need them? The answer is because infinity exists! If infinity didn't exist, we would be able to enumerate every possible case for any set of rules. It's easy, for example, to check that

2+2 = 4. Just get two apples, then get two more, and then see that there are now four apples. But, suppose that you wanted to say something about *every* possible addition. Well, there is an infinite number of them! And so, to say anything general, you need a proof that encompasses all infinite possibilities. But if, say, there were only a finite number of numbers, then you could enumerate every possible combination of numbers, and you wouldn't need a proof. *Proofs exist because infinity exists!* I think that's really cool!

2.6 The Tower of Knowledge

One issue is that to even get to Gödel, we need to build up a huge series of "modules". We need a module for mathematics, a module for humans, a module for philosophers, modules for Alan Turing and Alonzo Church and modules for computing, technology, electricity, mathematicians, formal logic, etc., *before* we can start talking about formal systems, incompleteness, and the equivalence (or lack thereof) between a human mind and a machine (such as a Turing machine). By the time we have all these modules, we are well past any possible realm of potential predictability, whether it be from the Practical Limits to Prediction (section *2, Determinism*) or from the Computational Irreducibility Limit to Prediction (section *2, Determinism*) or from the Chaos Limit to Prediction (section *3.14, Chaos*).

2.7 Substrate

Another way of thinking about the issue of FWIT is to think about the substrate. When you ask a computer for a "random number" what you get back is called a "pseudo-random-number". Its randomness is as good, or as bad, as the quality of the random number generation algorithm that is generating it. By definition, since the random number is being generated by an algorithm, it must be deterministic. And yet, you can produce random numbers from certain types of algorithms that pass known statistical tests for randomness. One example is the "Rule 30" cellular automaton shown in *3.14, Rules for evolving the "Rule 30" cellular automaton.*

> [...] Rule 30 generates seeming randomness despite the lack of anything that could reasonably be considered random input. Stephen Wolfram proposed using its center column as a pseudorandom number generator (PRNG); it passes many standard tests for randomness, and Wolfram previously used this rule in the Mathematica product for creating random integers. [154]

Wolfram Research's Mathematica software now has six different random number generating algorithms. One of them, "Extended CA",

> produces an extremely high level of randomness. It is so high that even using every single cell in output will give a stream of bits that passes many randomness tests, in spite of the obvious correlation between one cell and five previous ones.... In practice, using every fourth cell in each state vector proves to be sufficient to pass very stringent randomness tests. [165]

And yet, the substrate it is running on is completely deterministic. How can you get a nondeterministic outcome from that? You can't. *Free will is emergent phenomena running on a deterministic substrate in exactly the same way as a random number generator produces seemingly random numbers from a deterministic substrate.* Or, as Nobel Laureate Frank Wilczek says:

> Pseudo-randomness is a beautiful metaphor for how our own perception of free choice can emerge from underlying determinism as we navigate through the world. Just as

a sequence of pseudo-random numbers appears freely chosen if you don't have access to the program and seed, so can human actions appear to be freely chosen if you lack access to the brain's underlying subconscious processing. I suspect that this is more than a metaphor. [159]

Or, as Stephen Wolfram puts it:

But so in the end what makes us think that there is freedom in what a system does? In practice the main criterion seems to be that we cannot readily make predictions about the behavior of the system.

For certainly if we could, then this would show us that the behavior must be determined in a definite way, and so cannot be free. But at least with our normal methods of perception and analysis one typically needs rather simple behavior for us actually to be able to identify overall rules that let us make reasonable predictions about it. [161, pp. 751–752]

In other words, FWIP exists because we cannot predict.

By the way, as a complete aside, here is a fun fact. Once you realize that everything is deterministic, and you understand the concept of pseudo-randomness, you can immediately see that everything that seems *random* in our universe is actually *pseudo-random*![13]

2.8 Liability

If FWIT doesn't exist, then one big worry is that no one, not even a murderer or a thief, is responsible for their actions. After all, if what someone does is entirely determined by a physical law, then in what sense can they be responsible? However, this is not a good argument. There are several reasons why but by far the most important is that which was discussed in the preceding section *2.6, The Tower of Knowledge*. To talk about responsibility requires us to build an enormous tower on top of quantum field theory. We need modules for atoms, then molecules, then single celled organisms, then multicellular organisms, then mammals, then humans. We then need modules for ethics, justice, philosophy, law, morality, duty, and so on. Only after this exhaustive tower is built can we talk about "responsibility" or "liability". *A quantum field does not and **cannot** know what responsibility is.* That concept only makes sense after this entire tower of knowledge (or information if you prefer) is built— before that it makes *exactly* zero sense. Suppose you decide that you will not accept this argument. Then there is yet another major issue. While the "perpetrator" is following deterministic physical laws, *so is the rest of society!* So if you want to argue that the "perpetrator" is not responsible for his actions, then you will have to argue that society, which is also following the *same* physical laws, is somehow enjoined from stopping or punishing the perpetrator. But the situation is symmetric: what applies to the perpetrator applies to society as a whole too! You could take the position of an observer that is able to observe our universe from the outside: to make sense of what the set of quantum fields we call the "perpetrator" is doing, you will also have to understand what the set of fields called "society" is doing. And you cannot make sense of that *unless you understand the system of justice that the set of fields called "society" has created even if **all** the quantum fields involved are following deterministic physical laws.* You cannot get away from understanding liability, justice, and responsibility without understanding it at the level of *human* affairs, regardless of whether or not the underlying substrate is deterministic.

2.9 Levels of Explanation, Part 1

It is important to understand the issue of levels of scientific explanation. Because we cannot calculate the properties of one level from the adjacent level, for all intents and purposes, those levels are independent. That means that there are truly gigantic numbers of calculations that we cannot do. We can't even calculate the properties of an atom from quantum field theory, let alone collections of atoms, or molecules or anything bigger. And yet, we definitively know that all atoms must obey QFT, and even more specifically, all of chemistry and biochemistry must obey QED. The point is that free will (in practice) is "running at" a very, very high level.

And so we create functional explanations. A functional explanation is an encapsulation of a lot of complication from adjacent levels into one module that is easy to understand, digest, and use. That is why it is useful, not because it is "fundamental" in the deep sense of the word. Physicists do something similar all the time: they're called effective field theories. All the high energy behavior we either cannot calculate or do not want to deal with is encapsulated into some tractable mathematical object that works just fine as long as we work substantially below the energy scale whose behavior we are encapsulating. It works similarly with levels of explanation. When we aggregate a whole bunch of stuff together, we get a new object whose properties we have to discover since we cannot calculate them from the theory of the levels adjacent.[14] All of chemistry is contained in one equation, the equation of quantum electrodynamics. But we can't calculate much chemistry from QED except for extremely simple things. Similarly, all of biology is chemistry, but we can't get to the theory of evolution from chemistry because we cannot do the calculation. And so on. So what is an explanation? Let's say that Alice hurt Bob by whacking him with a baseball bat. One approach to explaining this is to think reductionistically. A portion of the quantum fields of the universe (Alice) interacted with another portion of the quantum fields of the universe (the baseball bat) and that combination then interacted with a third portion of the quantum fields of the universe (Bob), and all the portions of the quantum fields of the universe within their light cone during their interaction. At the level of quantum fields, all we will see (or be able to say) is that there were multiple portions of quantum fields, and they interacted in a certain way. But is that an explanation? That depends upon what it is that you want to do with that explanation. If all you want to do is predict what the configuration of the quantum fields would be after the interaction, you could (in principle) calculate the result of all these interactions, and you would know the new configuration. But that wouldn't tell you whether Bob was hurt, or why Alice whacked him with a baseball bat. For that, you would need a theory that treated Alice as a module, Bob as another module, a baseball bat as a third module, and a fourth module consisting of all those other modules (everything else within the light cone that didn't matter) that you would ignore. Entire towers of explanation are needed to create these three modules. *Only once those four modules existed could you say that you had satisfactorily explained if Bob was hurt or why Alice whacked him.* Crucially, without those towers of explanation, you could not create those modules, and without those modules, you could not explain any of this—in other words, you cannot *explain* human behavior by calculation from elementary quantum fields! And yet, human behavior *is* running on a completely deterministic substrate. To drive the point home, consider if you can tell if Bob is alive or dead after Alice hits him. To do that, you will need a theory of what is and isn't alive, and which field configurations yield a live person, and which field configurations yield a dead person. And you *cannot* figure out which field configurations yield which result without explanations at a level completely removed from the fields themselves. The same is true for a motivational explanation, or a chemical explanation, or psychological explanation or a biological explanation or an evolutionary explanation or an economic explanation—*any* explanation! They're all encapsulations of a gigantic amount of adjacent and lower-level behavior. They're all completely valid, very useful, and require enormous amounts of work and creative endeavor to produce. They cannot be produced by considering just the lowest level substrate. And yet they are all still running on, and composed entirely of, a deterministic substrate!

As an aside, consider the following. To even get to the point where we can argue about, say, Gödel (section *2.5, Gödel's Incompleteness*), we need a *gigantic* quantity of computational reducibility (see chapter 3, *Computation*) in the operating system of the universe (i.e., fundamental physics), and all of that reducibility better be deterministic, or we will be entirely unable to get from the lower-level laws to this point!

2.10 Modularity

It is the same pattern everywhere: in practice, to get anywhere, you have to modularize your understanding of heaps of stuff by treating each heap collectively as a "module" with known properties. That is what it means to understand something. Given this set of modules, what can I deduce about how a collection of them would behave? Computer scientists do not need to know the inner workings of a computer chip to be able to program it. They just need to know the parts that are exposed to them by the chip: its inner workings have been modularized to enable computer scientists to treat them as modules with some given properties. The computer scientist's job then is to figure out how to string those modules together to do something useful. Similarly, a chemist doesn't need to know the equations of quantum electrodynamics. It is more than sufficient to know that electrons live in "shells" around the nucleus of an atom, and those shells interact in certain prescribed ways to form molecules. An author of a book doesn't need to know how to program with a computer language: all they need to know is how to use a word processor. I do not know the internals of Mathematica at all, and I know very little about the internals of the typesetting language, TEX, and yet, I am using both of them extensively in writing this book. *Modularization means to hide the internals from the user!* In every case, the underlying knowledge is encapsulated, and that encapsulated knowledge is what is used to make further progress. An "explanation" then is showing how a collection of modules, plus their interaction rules, leads to a certain outcome. Let's try a statement in political science: money plus a dictator plus opportunity leads to war. Each of money, dictator, opportunity, and war are modules encapsulating unimaginably enormous quantities of underlying physical complexity. Or one from science: the Earth revolves around the Sun.[15] "Earth" is a gigantic collection of wave functions, "Sun" is a gigantic collection of wave functions and the space in between them is yet another gigantic collection of wave functions. Only if you treat them as "modules" can you come up with sensible explanations of what is going on.

2.11 Computational Equivalence

Cellular automata are sets of rules, some extremely simple, that when iterated can produce arbitrarily complex behavior. Yet, even when iterating the simplest rules, it is frequently the case that the only way to find out what a particular cellular automaton does is to run it. In other words, the very simplest rules can produce arbitrarily complex behavior to the point where, as Stephen Wolfram conjectured and Matthew Cook proved [34], certain cellular automata, despite their incredible simplicity, are universal computers![16]

The profundity of that outcome never ceases to amaze me, and I am surprised at how little thought this has engendered. For example, the Rule 110 cellular automaton [153] is known to be Turing complete: it can, in principle, run *any* computer program or do *any* calculation! Its output *cannot* be predicted — to see what it is all you can do is run it. What I am really arguing is that this cellular automaton is, in the most fundamental possible sense, entirely equivalent to *any* other universal computational system. It is equivalent to the workings of a human brain, or to the evolution of weather systems. All these things are computationally equivalent. [141] And if they are all computationally equivalent, then, if we have free will (in any sense) so does the weather and so does the Rule 110 cellular automaton. Therefore, it is much more productive to talk about free will as the inability to predict the outcome of a system without running the system all the way up to the point where the outcome occurs. *Free will is computationally irreducible!*[17] So, just as it makes sense to say things like "he has a mind of his own", it makes just as much sense to say "the weather has a mind of its own." For fun, let us compare the predictability of the weather, and the predictability of US Supreme Court justices (and I will assume that no one doubts that they have free will, at least in practice). Here is the predictability of the weather:

A seven-day forecast can accurately predict the weather about 80 percent of the time and a five-day forecast can accurately predict the weather approximately 90 percent of the time. However, a 10-day—or longer—forecast is only right about half the time. [108]

And here is the predictability of the vote of a US Supreme Court justice:

When put to the test, the model predicted like a champ, correctly returning a justice's vote 83 percent of the time. But the model truly proves itself when analyzing close cases where the judges are pretty evenly divided.

For three different scenarios, the researchers compared their models' predictive success with a "majority rule" method of prediction. Under majority rule, they would simply predict that the ninth judge's vote would follow the majority of the eight other justices whose votes are already known. If there was a tie, then the researchers would flip a coin to predict the final vote.

Majority rule actually works pretty well overall, correctly predicting a judge's decision roughly 70 percent of the time. When there's a sharp divide in the court (e.g. 5-4 decisions), however, majority rule fails miserably as a predictor.

The model shines under these scenarios. For instance, in 5-4 decisions, the model correctly predicted a judges' vote 77 percent of the time. Majority rule only made the right prediction 28 percent of the time. [11]

Yes, this is tongue in cheek, and yes, this is not meant to be taken particularly seriously. Having said that, the weather after seven days is about as predictable as the vote of a US Supreme Court justice!

All of this makes it clear that our "free will" is our own (and by extension everyone else's) inability to know what we're going to do until we do it because the workings of a human brain are computationally irreducible.

Some have argued that, at least in principle, you can simulate a human brain and run it much faster. However, how will we arrange that in silico simulation to get the input of every single interaction that our bodies have: every molecule of air, every smell, the light patterns when walking our dog, the interaction with people, etc.? Even the location of that in silico brain will be different. It may be arbitrarily similar to the brain of a specific person, but it cannot be that specific person in any recognizable sense of the phrase "specific person". So even that simulation will not be useful for exact prediction: it could certainly suffice for approximate prediction but never exact prediction.

2.12 Signal Compression

Consider a signal sent by Alice to Bob. The signal is 111 000 000 000 000 000 000 111 111 111 111. (Since any signal can be encoded as a binary signal, there is no loss of generality in using 1s and 0s.) Let's call this signal 1. How much information is in this signal, as compared to a second signal sent by Alice to Bob which is 110 101 101 000 111 011 010 011 010 100 001? Let's do a crude version of a procedure called "Run Length Encoding" (RLE) to see how much we can compress each signal by. In RLE, instead of sending the signal, you send a signal that contains the symbol (0 or 1) and the number of times it appears before it changes. So, for the first signal, you would send 1,3 0,24 1,12 (written in decimal rather than binary for clarity – it would obviously be more symbols in binary, but the point applies). Including the commas, instead of using 36 symbols (the spaces are for humans to be able to see this easily), you are using 11 symbols. Now let's try signal 2. Writing in decimal, we get 1,2 0,1 1,1 0,1 1,2 0,1 1,1 0,3 1,3 0,1 1,2 0,1 1,1 0,2 1,2 0,1 1,0 0,1 1,0 0,4 1,1. This yields 63 symbols (again more in binary), which is longer the original signal. Now it's true that RLE is likely not suited to the task of compressing the second signal—I chose it for its simplicity to illustrate the point. However, despite that, it's very clear that signal 1 is highly compressible, and signal 2 is very difficult to compress. Equivalently, signal 2 contains much more "information" than signal 1 because signal 1 is much more predictable. The more predictable a signal becomes, the less "information" it carries. And this is an excellent way of thinking about free will: the more predictable someone or something is, the less free will they have. In the limit I am discussing, since everything is deterministic there cannot be any free will in theory, but since nothing is predictable in practice, there is perfect free will in practice. Now, this doesn't stop us making approximate predictions. To the extent that we cannot even make approximate predictions about someone's behavior, that person is exhibiting more "free will" than someone who's behavior is less approximately unpredictable. And yes, I fully stand by the implication that, in some "fundamental" sense, those people that are more approximately unpredictable exhibit more "free will" than those that are more predictable.

2.13 Unitarity – Unitary Evolution

In the Schrödinger equation, what evolves is the wave function itself, conventionally denoted by the Greek letter "psi" but which we will simply call WF, which is a function of time, t. The wave function encodes all the information about the state of the physical system that we are considering. We use another postulate called the "Born rule" to define that the probability of a given observation is (a type of) squaring of this wave function. If we start with a wave function at time t_1, WF(t_1), we use the Schrödinger equation to evolve it to time t_2, where we get WF(t_2). If we square any (independent) possibility in the wave function, we get a number between 0 and 1: this is the probability of that event occurring. The key question is, what kind of evolution are we willing to allow? Must all probabilities be preserved? In other words, must all probabilities always add up to 1? The answer is yes. One of the most basic postulates of quantum mechanics is, in fact, "unitarity": unitarity *means* that all the probabilities sum up to 1 by postulate. This, together with the Born rule, guarantees that the sum of probabilities is always 1 — *information is **not** lost!*[18] This is a subtle point and needs to be clearly understood. The Schrödinger equation is time symmetric. You can evolve a wave function backwards from time t_2 to time t_1 or forwards from time t_1 to time t_2 (we are defining time t_1 to be less than t_2): neither Schrödinger nor his equation make any distinction between past and future. And in fact, nowhere in fundamental physics is any distinction made between past and future. The distinction between past and future is an *emergent property*: that distinction only emerges when we consider large collections of particles for reasons we don't need to get into here. Precisely because the Schrödinger equation, like everything else in fundamental physics, makes no distinction between past and future, you cannot make a "choice" in the Schrödinger equation, for a choice *requires* irreversibility, and the Schrödinger equation is reversible! (If you can always go back to the *exact same* configuration that you were in earlier, then in no sense could you have made a choice: that is *why* choice requires irreversibility.) In practice, the standard interpretation of quantum mechanics, the "Copenhagen interpretation", simply asserts that, at the moment a quantum system interacts with a "classical" system—a classical system, in this interpretation, is asserted to be a large enough system where quantum effects can be neglected, and the system can be treated "classically"—an irreversible choice occurs: this is the so-called "wave function" collapse (see section *2.14, Wave Function Collapse*). Only upon that "collapse" does a "choice" occur, which necessarily leads to information loss.

But without the "wave function" collapsing, no information is lost in quantum mechanics.

And, ignoring the possibility of "wave function collapse" for the moment, since information is not lost, it means that there is no "free choice" in quantum mechanics, so FWIT must be untrue. And the fact is, there is no "wave function collapse"—that's just a handy way of computing things in quantum mechanics: see section *2.14, Wave Function Collapse* for why. And in any case, as that section shows, "wave function collapse" doesn't help with the source of free will.

2.14 Wave Function Collapse

Quantum mechanics is, in and of itself, deterministic.

Quantum mechanics—as currently understood—is deterministic. The strange feature is what it determines: the PROBABILITY of what will happen. [61]

As distinct from the wave equation, which evolves deterministically, our knowledge of this evolution is constrained to probabilities only (the so-called "Measurement Problem in Quantum Mechanics"). Thus, to rescue FWIT, one would have to find some way of introducing nondeterminism into quantum mechanics. One way, seemingly very promising, to introduce nondeterminism into fundamental physics is via the "collapse" of the wave function. A wave function is a collection of probabilities:[19] what is the likelihood that x_1 will happen, what is the likelihood that x_2 will happen and so on, where the Xs are the entire collection of possibilities for what might happen.

Consider an electron, for instance. It is a "spin-1/2" particle. This means it has two spin states: spin up, and spin down. When you write the wave function of an electron, you write it as the sum of its spin up part and its spin down part. Algebraically, again denoting the wave function by WF, you say,

$$|WF| = a * |up| + b * |down|,$$

where the $|WF|$ is the total wave function, the $|up|$ is the up part of the wave function, the $|down|$ is the down part of the wave function, and a and b are the weights of each possibility (and when you square each of a and b and sum them you always get 1—this is called "unitarity"—see section 2.13, *Unitarity – Unitary Evolution*). But here is the puzzle: when you detect an electron, you *always* detect it either as having an up spin or having a down spin. You *never* detect the other possibility. So, the question is, *where did the other possibility go?* This is called the measurement problem in quantum mechanics.

To "address" this conundrum, Niels Bohr proposed what is now known as the "Copenhagen interpretation" of quantum mechanics. When you make a measurement in the "Copenhagen interpretation" you "collapse" the wave function so that only one of those possibilities is detected; the others just disappear into thin air. That disappearance of the other possibilities is nondeterministic: there are no physical laws governing that collapse.

> The Copenhagen interpretation of quantum mechanics is as easy to state as it is hard to swallow: when a quantum system is subjected to a measurement,[20] its wave function *collapses*. That is, the wave function goes instantaneously from describing a superposition of various possible observational outcomes to a completely different wave function, one that assigns 100 percent probability to the outcome that was actually measured, and 0 percent to anything else. [26, p. 239].

In terms of being able to calculate, the Copenhagen interpretation works very well. It always gives the right answer. It is known in the professional physics community as the "shut up and calculate" interpretation.

However, our purpose is very specific. We are trying to determine if the non-determinism inherent in wave function collapse can serve as the source of FWIT. So is this a promising way of introducing non-determinism into physics? No. Not at all. The error that leads to this conclusion is the assumption that the observer is not part of the quantum system. While that is a true enough assumption for the purpose of experimenting and trying to understand physics, it does not in any way tell us anything about the entire universe because by definition there is no way for the observer to be outside of the entire universe. In other words, the wave function of the universe also includes the observer and therefore it makes no sense to treat "that which is being observed" and the "observer" as separate.[21] The simplest way to think about quantum mechanics is to ask: what do the equations say? And then take that seriously as the correct description of reality. In the case of quantum mechanics, in particular, this is reasonable: quantum mechanics is the only physical theory where we have never, ever, got even an inkling of an experiment or observation that doesn't fit it. If we take quantum mechanics seriously, then what we are saying is that all probabilities are always preserved and all evolution is deterministic. There is no "wave function collapse" when making a measurement. That "collapse" is something outside of quantum mechanics and obeys none of its postulates, is non-unitary and so does not preserve probabilities, happens instantaneously, is completely nondeterministic, and has no mathematical or logical basis—it just "happens". And, as it turns out, is entirely unnecessary. Quantum mechanics works just fine without requiring the wave function to collapse: it produces the same results, the calculations are unaffected, and the theory is much simpler.

In other words, quantum mechanics is deterministic even though its output (observations) are probabilistic. That probability occurs because we only detect one of many possibilities when we make a detection. All the undetected possibilities are present in the quantum mechanical equations. But *only one of them manifests itself in a given observation at any given time*: all possibilities are present in the original equations. The probability of detection of a specific possibility is what we see when we *average out* our series of observations. All possibilities occur, each with a probability given by quantum mechanics; if a possibility is not in the quantum mechanical equations for the system we are considering, then that outcome *cannot* happen. Think of it this way. A coin flip is deterministic. Yet, all we can predict is that 50% of the time it will come up heads and 50% of the time it will come up tails. *Just because all you can predict is a probability doesn't mean that there is a lack of determinism!*

And so, for our purposes, we can skip all the standard, and completely correct, criticisms of the Copenhagen interpretation, and focus on just one criticism that is simply inarguable.

Once this objection is made, there is no further case to prosecute. And that objection is: the *detector* is made of the exact same things as the *detectee!* Every single part of the detector is following the exact same laws, *in every known respect*, as the detectee. The detector and the detectee are both part of the wave function of the universe. As humans, we arrange an "experiment" where we temporarily cause a portion of the wave function of the universe that is the detectee to "separate" itself from the rest of universe's wave function. If the universe's wave function is WF then we set things up so that WF is now equal to WF[Detectee] + WF[Rest Of The Universe Excluding Detectee]. We then separate out the wave function of the detector, so that we now get WF = WF[Detectee] + WF[Rest Of The Universe Excluding Detectee And Detector] + WF[Detector]. Our experiment then consists of seeing what happens when we let WF[Detectee] interact with WF[Detector] while ignoring WF[Rest Of The Universe Excluding Detectee And Detector]. Therefore, to say that the detector is able to, somehow, "collapse" the wave function of the detectee is a completely meaningless statement since it is *made of the same stuff* as the detectee. What is actually happening is that the wave function of the detectee is interacting with the wave function of the detector, where the wave function of the detector is some unimaginably complicated superposition of all the wave functions of all of its constituent parts, and the result is an even more unbelievably complicated superposition of the wave function of the detector *plus* the detectee. As a practical matter, it is incredibly difficult to show quantum mechanical effects on large collections of particles. Despite that difficulty, lately, real progress is being made in demonstrating these effects: "We have discovered an entirely new type of quantum phase transitions[sic] where entanglement takes place on the scale of many thousands of atoms instead of just in the microcosm of only a few," explains Vojta. [78] Therefore, there is no non-determinism there; just crazy complication making it hard, even in principle, to see how you would ever calculate the wave function of the detector, despite it indubitably existing. This means that you have to treat the detector as a "lump of stuff" rather than model each of its constituent parts if you want to get anywhere on a practical basis, which is why the Copenhagen interpretation works just fine for calculations. In any case, all the "collapse" does is pick out one of a list of possibilities so it can't be the source of free will: that list of possibilities is *not* under the control of the experimenter. So, going back to our electron example, if we repeatedly detect the spin state of an electron, the probability of finding an electron in a given spin state will be *exactly and always*[22] what quantum mechanics tells us it must be: there is no additional unpredictability or randomness there that could potentially be the source of free will. Therefore, none of this helps with free will in theory because there is no probability that is under the control of some observer (say a human) who can manipulate that probability to his/her own ends. Probabilities remain the same from one laboratory to another and from one scientist to another: if they didn't, then we wouldn't be able to discover scientific laws—we would just have a hodgepodge. *The very fact that different observers of the same experiment agree on the outcomes tells you that the "probabilities" present in quantum mechanics cannot be the source of free will in theory.*

2.15 Superdeterminism

There is a way to escape the inference of superluminal speeds and spooky action at a distance. But it involves absolute determinism in the universe, the complete absence of free will. Suppose the world is super-deterministic, with not just inanimate nature running on behind-the-scenes clockwork, but with our behavior, including our belief that we are free to choose to do one experiment rather than another, absolutely predetermined, including the "decision" by the experimenter to carry out one set of measurements rather than another, [then] the difficulty disappears. There is no need for a faster than light signal to tell particle A what measurement has been carried out on particle B, because the universe, including particle A, already "knows" what that measurement, and its outcome, will be. [74]

In other words, it's not that nature has rules that we can discover that tell us what it does. It is, rather, that every single thing that happens in the universe: every particle interaction, every smell, every itch, every thought, all motion and change—*everything that has ever happened or will happen*—is predetermined. This is superdeterminism and is a way of trying to escape from the fact that quantum mechanics *demands* that two particles that cannot communicate but are entangled at their creation will affect each other's state *instantaneously*. This makes some physicists (e.g. Einstein) uncomfortable: they call it "spooky action at a distance." For instance, suppose you created two particles at time t_0 and carefully separated them while preserving their state (by not allowing them to interact with their environment). Suppose that the two particles were electrons. Each electron *at the moment of its creation* is in a superposition of two states: spin up and spin down and you create them in such a way that their spins are "entangled"—completely locked together. Now suppose you separate them by the distance of a galaxy. The *instant* an experimenter detects electron 1 at time t_1 to be, say, "spin up" then that experimenter knows, with 100% certainty, that the other electron *will be* detected as having "spin down". This has now been shown to be true with every possible loophole plugged—see reference [29].

If the world were indeed superdeterministic, then it wouldn't need to follow the laws of physics! Unicorns could arise out of thin air, and on Tuesdays all the air over the Sahara could magically disappear leaving a vacuum and then magically reappear on Wednesdays. Because, under superdeterminism, since everything is completely determined at the beginning, no "choices" ever are made by anyone, *not even in practice!* Without any notion of choice, there is no notion of an experiment, and without any experiment there is no notion of a physical law. If there is no physical law, there is nothing to stop the world behaving in entirely random ways. Now, it's perfectly possible for the world to be *both* superdeterministic *and* follow physical laws. However, whether or not it follows physical laws is independent of whether or not it is superdeterministic. That independence, in fact, leads to a very subtle and unappreciated point that *requires* FWIT to be false *and* FWIP to be true: the probability of two independent things being true is always less than or equal to (and usually substantially less than) either one of them being true on their own. To put it another way: superdeterminism could just as easily make FWIP false but if FWIP is false, then all of science is completely false. And yet, we have not seen much evidence to suggest that science is false because the world is knowable and does follow rules that can be worked out. If FWIP were false then literally anything could happen—the laws of physics would not be a constraint, merely (and if even that) a guideline or suggestion.

2.16 Compatibility with a Deity

If the world is superdeterministic, then it means that everything is fully predetermined, and there are no physical laws (that is, the laws we see are an artifact, deliberately created or otherwise). There are *no laws* that particles follow—what they do is what was fully predetermined from the beginning. This position is incompatible with both FWIT *and* FWIP. No form of free will can exist in this scenario. By contrast, if the world is deterministic, this is perfectly compatible with a deity that sets the universe in motion according to fixed laws—the laws of physics—and then takes no further part in the proceedings. It is also possible that the universe is deterministic, exactly as we see, but that from time-to-time a/the deity intervenes in a manner contrary to physical laws. The trouble with that position is that there is simply no evidence for it whatsoever, despite the fact that such evidence would be completely trivial for a/the deity to provide. What do I mean by such evidence being trivial for a/the deity to provide? For example, a/the deity could simply part the Red Sea on every alternate Tuesday for 2 hours and then let the waters slowly return back to their original state. There would be no way of making any universal or believable law of physics that would be able to account for such a phenomenon. Or, a/the deity could randomly move a planet or moon from its orbit, every so often. Or a/the deity could stop the sun from shining for a few minutes, again at random.[23] There are an infinite number of things a/the deity could do to make it utterly clear and inarguable that it exists. Any of these examples would be definitive inarguable proof that a/the deity exists and intervenes in human affairs. The final possibility is of course that the world isn't deterministic in the first place. That leaves plenty of room for a deity involved in human affairs. But as of now there is no evidence for non-determinism in physics because, i) that is entirely contrary to our known knowledge of physics; and ii) there is no evidence for a/the deity's involvement in human affairs, despite the fact that this evidence would be even more trivial for a/the deity to provide than the previous examples. Do I really have to spell out such examples? OK, fine. A/the deity could provide a horse that spoke. Or could conjure up money out of thin air. Or could, at random, in full view of large numbers of people, levitate a giraffe and make it fly. Again, there are an absolutely infinite number of things which would be incredibly easy for a/the deity involved in human affairs to provide, which would invalidate the entire scientific enterprise of insisting that the world runs according to knowable physical laws and then discovering them, but instead runs according to the wishes and whims of a/the deity.

Furthermore, suppose you do believe in a/the deity that intervenes regularly in human affairs. Then you logically should believe in superdeterminism as well because then the world is overwhelmingly likely to be superdeterministic. This follows because you are effectively saying that the physical laws we see are artifacts, and do not describe reality. These "laws" are essentially coincidences resulting from how a/the deity organizes the affairs of the universe. Now you could argue that a/the deity intervenes very rarely indeed in human affairs, and not at all in scientific observations and experiments, and so the physical laws are real. But, in fact, that puts you right back into superdeterminism anyway: either everything that happens in the universe is governed by physical laws, or it isn't. If *everything* isn't, and there is even a *single* example of the universe not following (deterministic) physical laws, then the universe isn't deterministic! And yet we see these laws all over the place— therefore, the only possibility that would remain is that the universe is superdeterministic rather than deterministic. So if you are a deist, where a/the deity is of a type that intervenes in human affairs, you are also a superdeterminist. That is logically completely acceptable but will put you very far outside the scientific mainstream of thought. And it also means, *because it must*, that there is no free will whatsoever, either in theory or in practice.

If you believe in a/the deity that intervenes in the universe then you must believe in superdeterminism which means you cannot logically believe in free will, either in theory or in practice.

2.17 Missing Knowledge

Now, it could certainly be the case that there is more to physics than quantum field theory. And that missing part that we do not know is what supplies us with free will. This argument would be that fields do interact, nondeterministically, but by such an incredibly small amount that in standard particle theory experiments, we cannot detect their presence. But when we put a large quantity of quantum fields together, that property manifests itself as free will. This is logically consistent but then runs into the problem that, to every excited state of a quantum field,[24] we must ascribe some quantity of this new interaction. Therefore, every quantum field has within it a tiny bit of "free will" which when added up in large numbers, produces the free will we observe. This argument has the problem that we can see gigantic collections of quantum fields—the same quantum fields that we are made of—in things like planets, stars, and galaxies, which are many orders of magnitude larger than us and yet exhibit no free will whatsoever.[25]

2.18 Causality

The eminent physicist, Richard Muller, argues that "at the heart of physics, causality is lost" [102]. What he is arguing is that the collapse of the wave function is an example of the loss of causality, because "Quantum physics is completely causal—except for the missing part about the interpretation of the quantum state. The most successful 'interpretation' of that state is that it is non-causal when measured or detected." [102] In his book, however, he argues something different [103]: "Therefore, given complete knowledge of the past, even with accuracy sufficient to defeat chaos, certain important aspects of the future (such as one that might affect the lifetime of a cat) cannot be predicted. The most powerful historical argument against free will, the argument that formed the success of classical physics, the argument that physics was deterministic, was itself the illusion." Muller's argument is that we can have two identical particles in the laboratory and yet we cannot say which one will decay first: we can only give a probabilistic answer. The fact that we cannot give a definitive answer is, to Muller, proof that the future of these particles is not completely determined by their past. If the future is not entirely determined by the past then, according to Muller, that means that something more than just the wave function is affecting these particles. However, this ignores that fact that the wave function *is* deterministic and, *by definition*, encodes all possible knowledge about the state of any particle or physical system. All that is happening at the time of measurement is that the part of the universe's wave function that defines the particle is interacting with the part of the wave function of the universe that defines the detector. The experimenter and the laboratory, and everything else that it interacts with, constitutes a "detection". The result of the detection is the re-entanglement of all of those pieces of the wave function of the universe. And, belaboring the point, that is *exactly what the Schrödinger equation is saying!* All the other stuff is "cruft" added to the Schrödinger equation to try to make us feel better in some unspecified way: it is all superfluous, unnecessary, and entirely fails "Occam's razor". And so, Muller's argument, as I have shown and argued, is partly correct, and partly wrong. Physics *is* deterministic and to pretend otherwise is to shove one's head deep in the sand. But it is *precisely* the lack of predictability that is FWIP, and that lack of predictability comes not from something *outside* of deterministic physics, but rather as the *necessary consequence of it!* And in any case, as I have shown above in section *2.14, Wave Function Collapse*, Muller's argument doesn't solve the problem: the collapse of the wave function *cannot* be the source of free will because no "choice" is being made in the collapse of the wave function: even under the absurd Copenhagen interpretation, no observer is controlling the collapse—it simply "happens".

2.19 Levels of Explanation, Part 2
Nobel Laureate Philip Anderson argued that an explanation at one level doesn't explain things at another level.[26] For example,

all of chemistry is contained in one equation: the fundamental, quantum-field-theoretic equation of quantum electrodynamics. But that doesn't explain very much because we cannot calculate the properties of even very simple atoms using QED. However, the inability to mathematically derive the set of rules B from the set of rules A, does not, in any way, suggest that rule set B is nondeterministic while rule set A is. The *only* possible loophole is that there is some latent property in rule set A that is either unseen until there are lots of As interacting together, or is incredibly tiny and thus only manifests when many As are interacting together. This leads us right to the issue of: do rocks have free will? Rocks are made of the same stuff as we are and they too have lots of "As" interacting with each other, but we are still entirely unable to derive the properties of rocks from those "As". At the end of the day, rocks are indeed very much more predictable than humans. So where is the latent property in rocks that would endow them with free will?

One could argue that it isn't just a collection of interacting As that causes non-determinism, but that only certain configurations of interacting As cause nondeterminism. Even that fails. The laws of physics fully specify all the As, and all the interactions between all the As. Therefore, the only way to posit that certain configurations cause nondeterminism is a new interaction that is latent until many As are interacting in a specific configuration and/or a latent property.

But this too has a major problem. Does a freshly deceased body have free will? If the argument is that it doesn't, then we have a situation where, just before the body died it had free will, but just after it didn't. Where did it go? The infinitesimal change in configuration from alive to dead removed the free will? This is reminiscent of the religious argument of the existence of a nonphysical "soul". The problem is that there is no place in any physical law for a "soul" since all physical laws are deterministic.[27]

Let's think about this. Suppose you have an explanation at level X where the objects in questions are Xs. Level Υ is, say, one level above and it is made up of Xs, *and only* Xs. Let's further suppose that there are so many Xs that it is infeasible to calculate the behavior of Υ from that collection of Xs. Is there any doubt, however, that if every X is following deterministic laws, that a collection of Xs, no matter how large, is deterministic? Therefore, if Υ is made up of a collection of Xs, and only Xs, then it too must be deterministic. The only way you can get away from this determinism is *if and only if* some interaction occurs between those Xs as their number increases, *and* that interaction is *not deterministic* or if only special configurations produce nondeterminism or if there is a latent property.

At the most fundamental level there are only quantum fields. Excited states of those fields are called particles. Those particles interact with each other by exchanging other particles. Every one of those particles, because it is an excited state of a quantum field, is obeying a completely deterministic law. Every piece of a given person, every aspect of that person, fundamentally, is made up of quantum fields. *Unless an interaction can be found that is nondeterministic at a level above quantum fields it is a logical, mathematical, and physical impossibility for there to be a nondeterministic outcome.*

Furthermore, we are made up of the same things—i.e., the same atoms—as rocks, water, sodium chloride, hydrogen, etc. So do rocks have "human-like" free will? What about a jar of hydrogen? What about "empty space"?[28] There is about one atom per cubic centimeter in space. Does that cubic centimeter have free will? Can it feel? Does it have consciousness too, but just a little bit? There is a famous theorem in quantum mechanics due to Conway and Kochen called the "Free Will Theorem". [33] It shows that if three, very simple things (which no one doubts) hold true, then *if an experimenter has free will, then so does the particle that the experimenter is detecting.* They call those three things the SPIN, TWIN and FIN axioms. For our purposes, we can ignore their details but just focus on the result. Conway and Kochen argue that the result of their theorem shows *either* that the world is entirely deterministic *or* that individual particles have free will. My point, and the point in this entire book is this: you are free to believe that individual particles have free will; that is, that their actions are *not* fully determined by the state of the world at any given time. However, that is an incredibly extreme position to take when it is much easier and simpler to realize, yet again, that *free will is the inability to predict what you are going to do before you do it.*

And why can't you predict? Because to accurately predict, you would have to simulate every single interaction between you and the environment—every photon of light, every molecule of air, every atom of food etc. that you will interact with between now and prediction time. And, while keeping track of all of that, you will have to simultaneously model your own internal state with perfect accuracy: every molecule in your own body. And you must do that in a way that avoids the model interacting with the real you! This requires that you distance yourself by 300 million meters for each second that the experiment continues. The experiment must then somehow indelibly record its results so you can see them upon your return! If we traveled exactly at the speed of light, you would return simultaneously with the outcome of your experiment. Anything slower gets you there after the fact—and so you see the result on yourself before you see the simulation! There is no way to get the simulated answer first! Only if you ran this simulation and could somehow get the answer faster than the speed of light (which is forbidden by well tested physical laws) would you be able to predict what you are going to do before you do it. As long as you cannot predict what you are going to do before you do it, you have free will. You have to *actually make the choice to know what choice you have made!*[29]

And this is all even before thinking about computational irreducibility (see section *3.15, Irreducibility versus Chaos*). A computationally irreducible computation must be carried out to know its result—there is no shortcut to get to the result without going through all the intervening steps. And that is free will— *you have to do what you have to do to know what it is that you did!*

As Stephen Wolfram notes [emphasis added],

When viewed in computational terms most of the great historical triumphs of theoretical science turn out to be remarkably similar in their basic character. *For at some level almost all of them are based on finding ways to reduce the amount of computational work that has to be done in order to predict how some particular system will behave.*

Most of the time the idea is to derive a mathematical formula that allows one to determine what the outcome of the evolution of the system will be without explicitly having to trace its steps.

And thus, for example, an early triumph of theoretical science was the derivation of a formula for the position of a single idealized planet orbiting a star. For given this formula one can just plug in numbers to work out where the planet will be at any point in the future, without ever explicitly having to trace the steps in its motion. [161, p. 737]

2.20 Bell's Theorem

Some have argued [51] that because Bell's theorem shows that there are no viable *local* theories of quantum mechanics, there is indeterminism in quantum mechanics and this indeterminism is what enables free will. However, this argument is entirely false. Non-locality simply means that, via entanglement, the result of an experiment (usually when the wave function of the experimental apparatus—i.e., the detector and detectee—extends over a large distance) can be affected by events that occur outside its light cone.[30] It *emphatically does not* in any way mean that those effects are somehow nondeterministic. At the end of the day, the entire experiment such as it is—the detector, the particles, the experimenter, and all the space intervening are part of one large wave function. That spatially separated parts of this wave function, outside each other's light cone, affect each other, means nothing more than "there be non-locality"! The existence of the overall wave function (which is, of course, incalculably complicated) itself refutes the existence of nondeterminism. Non-local does not mean nondeterministic!

Interestingly, this non-locality is of a very specific kind. Despite the fact that events from outside the experiment's light cone can affect the results that the experimenter gets (which is Bell's theorem) *this non-locality cannot be used to transmit any kind of information.* And because it cannot be used to transmit any kind of information, *it cannot be the source of free will.*

2.21 What's an Experiment?

What do I mean by an experiment then? It means that we have "made" a section of the wave function of the universe particularly simple to help us deduce something useful. And similarly, an observation (such as an astronomical observation) means that a section of the wave function of the universe is sufficiently simple to be "understood" or "observed". Furthermore, the experimenter really is making a choice in the experiment via FWIP (if we rule out superdeterminism. I find the argument in section *2.15, Superdeterminism* above compelling and so rule out superdeterminism). If things are deterministic but not super deterministic, then the laws of physics really do exist. Therefore, an experiment is the action of creating a situation that is sufficiently simple such that the laws of physics can be deduced. The experimenter is trying to extract the rules for the underlying regularity and to do that he has to simplify the system(s) under consideration as much as possible. So the experimenter does make choices (in practice, and only in practice!) but those choices are themselves constrained by the laws of physics—that is, the experimenter could not make a choice that the physical laws do not allow. So, for example, the experimenter couldn't decide to run an experiment where the mass of the proton was equal to the mass of the electron or where the photon was massive—the laws of physics do not allow that.

2.22 Self Reference

If you knew what you were going to do, then you might change it. And that creates a self-referential problem. Consider a closed loop in general relativity: what would happen if you could go back in time and stop your parents from meeting? The outcome is inconsistent. Similarly, if you knew what you were going to do, you could then not do it, but then if you didn't do it, then how did you know that you were going to do it? Once again, the lack of the ability to predict solves the paradox: since you cannot know what you are going to do, despite the system being entirely deterministic, this problem doesn't obtain. Surprisingly, at least to me, this *doesn't* apply to Epimenides' paradox which asks us to think about the statement: "All Cretans are liars". Is it possible for Epimenides, himself a Cretan, to have spoken the truth? In this case, however, the paradox can be resolved by assuming that the statement is false. No contradiction is created in that case. See [146] for example. I find that astonishing, but I can't explain why!

2.23 A Paradox

One of the issues that has always bothered me about fundamental physics is the following question: if we are part of the universe, and we wish to find a "theory of everything", then how is it that *a part of the universe can understand the whole?* To me, this creates a self-reference type problem—is the universe simulating itself? How can that be? Interestingly, the deterministic paradigm produces a very nice and clean answer to this question. The key point is that it is very normal for the description of the *part* to be more complicated than the description of the *whole*. And, in fact, the part can be arbitrarily more complicated than the whole. The whole could have a very simple description, but then trying to describe a part of it could be (and frequently is) incredibly complicated. One example of this is provided by cellular automata, like the rule 110 cellular automaton of section 3.5, *Computational Equivalence*, where very simple rules produce output that is capable of being a universal computer. The rules are the "description of the whole" while the output is the "part": the formula for the whole is very simple—just the stated rules—but trying to write a formula for any given part will be arbitrarily difficult. As Max Tegmark points out [132]:

> But an entire ensemble is often much simpler than one of its members. This principle can be stated more formally using the notion of algorithmic information content. The algorithmic information content in a number is, roughly speaking, the length of the shortest computer program that will produce that number as output. For example, consider the set of all integers. Which is simpler, the whole set or just one number? Naively, you might think that a single number is simpler, but the entire set can be generated by quite a trivial computer program, whereas a single number can be hugely long. Therefore, the whole set is actually simpler.
>
> Similarly, the set of all solutions to Einstein's field equations[31] is simpler than a specific solution. The former is described by a few equations, whereas the latter requires the specification of vast amounts of initial data

on some hypersurface. The lesson is that complexity increases when we restrict our attention to one particular element in an ensemble, thereby losing the symmetry and simplicity that were inherent in the totality of all the elements taken together.

Which solves the paradox. Since "simulating the whole" is much simpler than "simulating a part", it's quite reasonable to expect that a part of the universe can find simple rules to describe the "entire universe". One really cool thing to note is the following. Suppose you flip around the above and instead of thinking about the whole universe, think about the universe as a sum of its parts. We already know that the description of the whole universe is simpler than that of its parts. And what that means is that the *complexity in each of its various parts, no matter how you divide up the universe, all cancels leaving behind the simplicity of the whole!* I find that stunningly beautiful.

And, interestingly, this also "solves" the seeming paradox of consciousness. Because, at the most fundamental level, consciousness means that the conscious being is thinking about itself. So how can a part of a human brain think about the "whole" of the brain's experiences? Well, for one thing, since the vast majority of our experience is unconscious, it doesn't. But nevertheless, the reason that a part of the brain can wonder about the "whole" is precisely because a part can be arbitrarily more complex than the whole—it doesn't *have* to be, but it certainly *can*. And that part can then definitely think about the experience of all the other parts of the brain (or body for that matter!).

2.24 Provisionality & Paradigms

One common objection to determinism is that scientific theories are held to be "provisional" and incomplete because it is always possible for new evidence to overthrow the previous paradigm.

How do we know that some nondeterministic theory won't come along in the future and overthrow the deterministic paradigm? This objection is ill-founded with respect to fundamental physics. Physicists discover equations to explain behavior at different "energy scales". Different phenomena at different energy scales have different equations. For the energy scales at which a theory is written and tested, i.e. the energy scales at which the theory is valid, *that theory is correct and will always be correct!* Newtonian mechanics was supplanted by Einsteinian general relativity. Yet, we use Newtonian mechanics all the time and it works. At the energy scales at which Newtonian mechanics is valid it will *always be valid.* In fact, a good test of any new proposed theory is always: *does this new theory reduce to the old one at the energy scales at which the old one is valid?* Because if the new theory *does not* reduce to the old one at the appropriate energy scale, we can reject it: it is *false!* In other words, unlike knowledge in fields like accounting, politics, economics, anthropology, biology, etc., knowledge in physics is forever. Once you discover an electron, you've discovered it. No future theory can make the electron no longer exist and still be correct!

Next, all of biology is driven by chemistry: organic chemistry to be exact. Humans are biological creatures made up of organic molecules and (as I said earlier) those organic molecules are made up of the usual elements. *All of chemistry reduces to one equation from physics:* the equation of quantum electrodynamics (the theory of light and ordinary matter). Nobel Laureate Paul Dirac, one of the founders of quantum theory, once colorfully said:

> The underlying physical laws necessary for the mathematical theory of a large part of physics and the whole of chemistry are thus completely known, and the difficulty is only that the exact application of these laws leads to equations much too complicated to be

soluble. It therefore becomes desirable that approximate practical methods of applying quantum mechanics should be developed, which can lead to an explanation of the main features of complex atomic systems without too much computation. [100]

Dirac also said:

Quantum mechanics has explained all of chemistry and most of physics. [100]

And this is true! And it will remain true. In other words, *the laws of physics that describe all of human existence are already known!* Let that sink in for a minute. The laws that underlie all of human existence, every smell, taste and touch, every interaction, every stock purchase, every virus, pandemics, jungles, baboons, rain, snow, mountains, whales, beetles, etc., *are already known!*

We know that this must be true. Why? Because we know that ordinary matter — the matter from which we are made — has remained the same for *billions* of years. An Oxygen atom here on earth is *exactly* the same as an Oxygen atom far across the universe. And not only that, an Oxygen atom here on earth today is the same as an Oxygen atom was billions of years ago.

And we know this because our scientific instruments tell us that. Our instruments are now so exquisitely sensitive that we can detect these elements in exoplanets hundreds of light years away. How do we do that? The answer is because of "spectra".[32] Each element makes up its own pattern of "dark lines" in the light we detect. So by looking at distant spectra, not only can we see things far away, but we also see them as they were in the past. On Christmas Day, December 25, 2021, NASA launched the James Webb Space Telescope (JWST), which instantly became the largest optical telescope in space. It is exquisitely sensitive to infrared light which enables it to see light that has been redshifted by its multi-billion-year traverse of space. It also enables it to see the signatures of carbon dioxide and water vapor. On July 10, 2022, astronomers looked at WASP-39b, an exoplanet about 700 light years away. JWST detected the

> first clear evidence for carbon dioxide in the atmosphere
> of a planet outside the solar system.... The research team
> used Webb's Near-Infrared Spectrograph (NIRSpec) for
> its observations of WASP-39b. In the resulting spectrum
> of the exoplanet's atmosphere, a small hill between 4.1
> and 4.6 microns presents the first clear, detailed evidence
> for carbon dioxide ever detected in a planet outside the
> solar system. [106]

That's the same carbon dioxide 700 light years away as we have here on earth. And that means that it's the same carbon atoms and the same oxygen atoms, and they are bound together in the same way as here on earth. And that means that the quantum field theory of the subatomic particles that make up Carbon and Oxygen is exactly the same 700 light years from earth as it is on earth.

But there is still more! We can detect the same elements in distant galaxies with light that has taken billions of years to reach us. When we look at a galaxy far away, we are seeing it as it was when the light we detect left it. With our new instruments, such as the JWST, we can look back almost to the beginning of time—that is, almost 13.6 billion years ago when our universe came into being. We can detect the signature of Oxygen (and every other element) in the light emitted by galaxies that was emitted billions of years ago. And that pattern for Oxygen (and every other element) hasn't changed despite the fact that we are detecting it from billions of light years away and despite the billions of years that have passed since that light was emitted.

So when you read about a scientific experiment, such as the Large Hadron Collider, making such and such discovery or finding such and such discrepancy in the "standard model of particle physics," you must realize that it is operating at energy scales incomprehensibly far above human existence.

> The standard model predicts that at very high temperatures combined with very high densities, quarks and gluons would exist freely in a new state of matter known as quark-gluon plasma, a hot, dense "soup" of quarks and gluons. Such a transition should occur when the temperature exceeds around 2000 billion degrees—about 100000 times hotter than the core of the Sun. For a few millionths of a second, about 10^{-6} s[econds] after the Big Bang, the temperature and density of the Universe were indeed high enough for the entire Universe to have been in a state of quark-gluon plasma. *The ALICE experiment will recreate these conditions within the volume of an atomic nucleus, and analyse the resulting traces in detail to test the existence of the plasma and study its characteristics.* [emphasis added] [85]

Similarly, our discoveries of dark matter and dark energy, both of which we still do not understand, have no effect whatsoever on our biology.[33] So as we discover newer and newer features about higher and higher energy scales, our knowledge of the universe will improve. But it will not change the knowledge we already have about the physics of human existence. And that knowledge, those known laws of physics that apply to biology, are entirely deterministic now, and will remain entirely deterministic in the future. No future discoveries are going to affect what we already know at the energy scale of our existence.

2.25 Cockroaches

Let's consider to what extent a cockroach has free will in practice. It's certainly true that it is a simple creature and tends to follow relatively simple rules. Cockroaches have a behavior called "wall following": if you touch one antenna, it turns in the opposite direction. However, Greg Gage invented the cyberroach. And this invention shows both how "physical" a brain is: that is, how it is essentially a computer running an algorithm and, simultaneously, it shows the existence of free will in practice! Gage surgically fitted an electronic backpack onto a cockroach. That backpack sends a current into the nerves of the cockroach's antennae. When you stimulate those nerves with an electric current, the cockroach responds by turning. The cockroach follows a simple algorithm: if its right antenna touches something, it turns left, and vice versa, hence "wall-following". So here we are: we have an animal following a simple algorithm, and we can control that algorithm by means of an electrical impulse. Entirely physical: no metaphysics involved whatsoever. But then it gets very interesting: "But very soon, it adapts to a[n] unnatural stimulus. After 15 minutes, the roach ignores the backpack. It retains its free will." [52] So we have a simple creature, with a simple brain, but yet like bees (see section *3.8, Minimality*), it's capable of very complex behavior. The behavior is so complex that it can *ignore a stimulus that it knows is incorrect!* It is absolutely exhibiting free will in practice!

2.26 DNA

The plummeting price of genetic sequencing, the rise of AI, and the beginnings of an understanding of the biology behind mental pathologies is making it increasingly clear that humans are a biological machine made of the same parts (atoms) as every other thing in the universe—animate or inanimate. The most fundamental aspect of this is the information that is used to "grow" an organism—i.e., DNA—the "recipe" or "program" for how to build each living creature. DNA is made of chemical building blocks called nucleotides. Each nucleotide is made of three parts: a phosphate group, a sugar group, and a nitrogen base. There are four types of nitrogen bases: adenine (A), guanine (G), cytosine (C) and thymine (T). The order of these bases is a code. That code, i.e. DNA, encodes all the information required to build and operate a human, or any other living creature. The purpose of the code is to encode the instructions for making proteins, which are the molecules that do most of the work in our bodies. A sequence of those bases, which ranges in length from a thousand to a million, is called a gene. Each gene encodes the instructions for making a single protein. Genes make up around 1% of the DNA sequence: the rest of the sequence is used to regulate when, how, and how much of a given protein is made. There are roughly 3 billion bases and more than 99% of those bases are the same from one human to another. So what makes us different from each other is the difference in the information encoded in those bases, and our life experiences from conception to death encoded as information in our bodies (mainly our brains). Drilling down further, phosphate groups are made of a phosphorus atom bonded to four oxygen atoms, sugar groups are made of carbon, oxygen and hydrogen bonded together, and nitrogen bases are made of nitrogen, carbon, and oxygen bonded together. The entire code for all living things on earth is stored in DNA. And that code is made up of the very same stuff as the rest of the universe—from a biological point of view the *only* interesting thing about these atoms *is their arrangement!*

Notes

1. Of course, we are talking about exact prediction, not approximate prediction.
2. Even the very simplest exact calculations in quantum mechanics are incredibly difficult. Take a gas for instance.

Once it has more than just 40 particles, it becomes impossible for even the biggest supercomputers to use quantum mechanics to calculate anything. This is because the difficulty increases exponentially with the number of particles. Recently, there have been attempts to use Machine Learning to help make predictions in situations where direct calculation is impossible. Of necessity, such predictions are approximate, not exact. To the extent that those approximations suffice for predicting human behavior (there is no sign that they do, as of yet), it will naturally shrink the domain of Free Will in Practice. [64]

3. I discuss this in section 3.15, *Irreducibility versus Chaos*.

4. In fact, this understates the potential benefit. For example, for older adults with pets, "over 6 years, cognitive scores declined at a lower rate in pet owners than in people without pets." [58]

5. In fact, we *use* the constancy of the speed of light (in a vacuum) to *define* the meter. The meter is *defined* as the length of the path traveled by light during the time interval of 1/299,792,458 of a second.

6. For example, see reference [54]. As Nick Lane puts it: "Life needs a far-from-equilibrium environment, in which the forward reactions of metabolism are sustained by continuously reactive surroundings, most simply by continuous flow." [86, p. 154]

7. This is not quite as crazy as it sounds. Many countries have had 90%+ income tax rates—usually, with all kinds of loopholes—in the past. This includes the US: in 1944–45 the highest marginal tax rate was 94%.

8. If you do a web search for "Laffer curve" you'll find all sorts of proposed shapes. I'm not taking any position on what its shape is. My point is that it is *constrained* by its two fixed points, and therefore must bend back on itself. It is that bending back on itself that causes all the conceptual difficulty.

9. This is why we give the government a monopoly on committing violence against its citizens. Some entity must be able to force contracts to be honored and taxes to be paid. It is either the "law of the jungle" where you pay your tribute to the biggest, strongest, and most violent thus hurting the weaker members of a society, or a "civilized society" where all have equal rights and the state (i.e., the violence monopolist) enforces those rights.

10. See, for example, these four references, all by N. Gregory Mankiw: [90] [25] [91] [92].

11. Arthur Okun, the top economic advisor to President Lyndon B. Johnson, wrote an (excellent) entire book on this subject: *Equality and Efficiency: The Big Tradeoff.* [113] See Roger Lowenstein's article [89] in *The Wall Street Journal* for an excellent summary.

12. Tyler Cowen has an excellent book, *Stubborn Attachments,* on this topic. [38]

13. With the exception of the collapse of the wave function if you are a supporter of the Copenhagen interpretation of quantum mechanics.

14. Almost certainly due to computational irreducibility.

15. Really, they both revolve around their common center of mass, but since this center of mass is deep within the sun, we can ignore this slight deviation. It doesn't change the argument anyway.

16. A universal computer can run *any* possible program.

17. See section *3.15, Irreducibility versus Chaos.*

18. See a wonderful article in *Quanta* by Charlie Wood, "Physicists Rewrite a Quantum Rule That Clashes With Our Universe", for more explanation of Unitarity and how physicists are trying to reconcile unitarity with an expanding universe—something we must do if we are to have a theory of quantum gravity. [166]

19. Actually probability amplitudes, but that isn't important for this discussion.

20. Exactly what constitutes a measurement is undefined in this insane interpretation.

21. Even as a piece of just "physics," the Copenhagen interpretation is complete nonsense. It is the worst kind of science. In Wolfgang Pauli's famous phrase, "It is not even wrong!"

22. Plus or minus experimental errors and statistical fluctuations.

23. In fact, this is why pre-scientific man was so scared of solar eclipses. Until we understood them scientifically, there seemed to be no rational explanation for them other than the gods were angry!

24. A particle is an excited state of a quantum field.

25. Because they obey Newton/Einstein's gravitational laws.

26. See, for example, an excellent obituary in reference [31].

27. Superdeterminism may potentially allow for a "soul" but it would depend upon what the definition of a soul is. See section 2.15, *Superdeterminism* for an explanation of superdeterminism.

28. Space isn't really empty because there are quantum fields popping in and out of the vacuum but this doesn't change the argument.

29. Or to be political for a moment, in the then Speaker of the US House of Representatives Nancy Pelosi's famous words, "We have to pass the bill so you can find out what's in it"!

30. See the definition of light cone in section 1.1, *The Argument*.

31. The equations for general relativity.

32. Jonathan McDowell has a really nice tweetstorm explanation of why spectra get astronomers excited. See reference [95].

33. In fact, dark matter and dark energy *cannot* affect our biology or chemistry in any way. Whatever they are, and in the case of dark matter if it even exists, we know that they only interact with the ordinary matter of which we are made via gravity. Our bodies weigh on the order of about 100 kilograms, barring factors of 2. 100 kilograms of any matter, dark or otherwise, has such an incredibly small gravitational effect, that there is no possibility of its gravity affecting our biology. And there is an exponentially smaller gravitational effect of the dark energy that may surround our body. Dark energy has an effect on galactic scales. Dark energy has an effect on universe sized scales! Neither is relevant to our biology or chemistry. We evolved in the *earth's* gravity—that has a gigantic effect on us. But *our own* gravity is too small to be measurable or relevant.

Chapter 3

Computation

My argument about free will is not dependent upon thinking within the computational paradigm. Quantum Theory (QT) made up of Quantum Field Theory (QFT), Quantum Mechanics (QM), and General Relativity (GR) all stand on their own, and all are deterministic. The world we live in is entirely described by these 3 theories. However, in this section, I am writing as if the computational paradigm applies to physics and thus by extension to ourselves and all of the universe.

3.1 Physics and Discreteness

There is a (small) fly in the ointment when doing this. QT shows us that energy is discrete: it comes in little quanta that are multiples of Planck's constant. You cannot have an energy that is not a multiple of Planck's constant—in other words, energy is discrete. On the other hand, in QM as well as GR, space and time are treated as continuous. But computing, of necessity, is discrete. So the question becomes: is thinking about physics via the computational paradigm an approximation to the underlying continuous quantities? Or is it the case that the continuity of space and time in QT is actually an emergent property when you aggregate discrete bits of space and time?

There is no provable and definitive answer to this question for now. However, we have extremely strong hints that space and time must be discrete. For one thing, we do not have a fully complete quantum theory of gravity. When the gravitational field is not very strong, we can use "semi classical quantum gravity" where we treat the matter fields as being quantum, and the gravitational field (i.e., space and time) as being classical. However, this doesn't work when the gravitational field becomes extremely strong, such as on the edge of a black hole. From current work in quantum gravity, it is becoming increasingly clear that there is unlikely to be a synthesis of QT & GR unless *everything* is quantized (i.e., discrete), including space & time. There have also been explicitly computational attempts at quantum gravity, for example reference [88]. Evidently, way back at the beginning of quantum theory, Albert Einstein himself expected that as our description of the universe gets better, it will of necessity become discrete. [70] If everything really is discrete then, by definition, the computing paradigm applies (because it is not just an approximation) and therefore can help us think through a workable model of free will.

3.2 Bits

Why do I keep saying that all that matters is the arrangement, and only the arrangement, of atoms and subatomic particles? Well, consider a bit. The basic unit of computation, a bit, has a value of either 0 or 1. It can have no other value. Obviously, to compute with bits, we must store them. If we can't store them, we can't compute with them. To store bits to use them, it means we *must* physically instantiate them in some matter. For example, take DRAM (dynamic random access memory): this is the standard computer memory that our computers use to do their computations. It consists of a transistor and a capacitor. When we want to store a "1", we charge the capacitor. When we want to store a "0", we discharge the capacitor. Charging a capacitor is a physical act: electrons are moved from one side of the capacitor to the other so that there is an excess of electrons on one side, and a deficit on the other. That is what it means for the capacitor to be charged. The electron is, of course, a *subatomic* particle. In other words, modern computation relies upon subatomic particles! In a similar fashion, an idea or memory or a thought in our brain consists of the "firing" of neurons. That firing also consists of moving around subatomic particles (i.e., electrons and protons). When we learn, our brain physically changes. For example, on a PET scan, the brain of a musician looks different than the brain of a non-musician. All that musical training has physically changed the brain. The matter that makes up the brain *changes* as it works. With new techniques in "ultrafast ultrasound" we can actually see the matter of the brain changing as it works: "Ultrafast ultrasound could trace brain signaling with great precision, documenting how circuits and groups of cells interact as the brain performs functions from perception to decision-making." [121] To do anything at all, a physical change in the arrangement of the matter of the brain must take place. *Thoughts, and any other mental activity, are the physical act of rearranging the material of which the brain is made!*

This also leads to an interesting bit of metaphysics, unrelated to our main theme. The platonic idea of "perfect" things that exist out there independent of our thoughts is very likely to be incorrect. In fact, this idea can cause all kinds of practical mischief as I argued in section *1.7, Physicalism Denial*. All information, whether it be thoughts, mathematics, philosophy, or anything else, does not exist until it is "instantiated" in some physical object. Without that instantiation, there is no way for it to exist: there is no information unless something exists within which the information can be stored and/or processed. *Mathematics exists because the physical world exists — it cannot exist without the physical world!*

3.3 Reversibility

A reversible system *cannot* make a choice! Since the system is reversible, it means that whatever its state may be at some time, you can always wind it *back* to any state it had in its past *without the expenditure of energy!*[1] That means that every future state must be uniquely reachable from a specific past state because, if there is a future state that is reachable via more than one past state, information is lost—you cannot wind back the clock to the past state anymore. That is, in fact, what *reversibility* means. A reversible process does *not* have an arrow of time. For there to be an arrow of time, you *must* have irreversibility. And you cannot make a choice without an arrow of time—i.e., you must have irreversibility. But all fundamental physical laws are reversible! So, if the underlying system (i.e., governed by the laws of physics) is reversible, then how can we make a choice? The laws of fundamental physics look exactly the same going forwards as backwards.[2] So what gives? Well, if you think in computational terms (as Wolfram has pointed out) there is no mystery. Time is merely the inexorable grinding of the next computation that computes the new state of the universe from the previous state of the universe. And that, plus computational irreducibility,[3] is the ultimate source of Free Will in Practice in the Universe under the computational paradigm.

3.4 Universality: Definition

Suppose we see a complicated result. Let's say we see a complicated pattern of square tiles on a white background. Our intuition would suggest to us that the more complicated the pattern, the more complicated must be the rules that create it. However, it turns out that a few incredibly simple rules can produce arbitrarily complicated output. Furthermore, it turns out that not only do these simple rules produce very complicated output but also that many different such simple rules all produce equally complicated output. And, still further, it turns out that the complicated outputs produced by those simple rules are exactly *as complicated as* any outputs from much more complex sets of rules. This leads to a key observation: *extremely simple rules can be universal: in other words, there exists an encoding such that we can use even incredibly simple rules to do any possible computation!* There are a lot of possible universal computing systems. Some examples are DNA, Turing machines

[158], lambda calculus [122], one instruction set computers,[4] the S and K operators of Moses Schönfinkel,[5] and cellular automata.[6] Each of these (other than DNA) has incredibly simple rules— they are almost trivial. And yet, these almost trivial sets of rules can produce behavior that is universal!

So, what do we mean by universality? This follows from something called the "Church-Turing" thesis, named after the American mathematician Alonzo Church, and the famous (and persecuted) British mathematician, code breaker and computer scientist, Alan Turing. There are a lot of subtle and complicated issues with regard to this thesis which are outside the scope of this book. For our purposes, it is enough to understand that the Church-Turing thesis is about the nature of "computable functions" and says that any such "computable function" can be computed by a Turing machine. Before Turing, mathematicians used the term "effectively calculable" to mean a function that, *at least in principle,* is capable of being calculated via pencil and paper methods, and thus by a Turing machine. The two terms "computable function" and "effectively calculable" are, for our purposes, synonyms. The subtlety is that, while the Church-Turing thesis has almost universal acceptance, it is not possible to prove it, because there is no formal definition of "effectively calculable": mathematicians use an informal definition that is good enough for all practical purposes. For our purposes, we can say that a function (for example, from numbers to numbers) is "computable" if and only if it can be calculated via a Turing machine. *No one has yet found an example of a function that is not calculable by a Turing machine but is calculable by some other device or method.* Therefore, we *declare* that a Turing machine is universal. This means that universality can be defined by comparison to a Turing machine. Consider some other method or machine (such as the examples above) which we can call M. Therefore, M is universal if there exists an encoding that translates every computation done on M to a computation on a Turing machine. There are subtleties here too. As Wolfram puts it:

> Like many fundamental concepts, universality is in some ways clear and straightforward to define, and in other ways almost arbitrarily difficult to pin down. The general idea is that a system is universal if it can be "programmed" to emulate any other system. The difficulty comes in understanding what kinds of programming should be allowed. For if too much goes into the construction of the program, then computation can be done there, rather than in the system itself.
>
> One common definition requires that the encoding of input for the universal system, and the decoding of its output, must both be done by a computation that always halts after a finite number of steps. One difficulty with this definition is that it may be too restrictive for infinite tapes....
>
> In most cases, it should nevertheless be fairly obvious whether something should be considered a valid encoding for a universal system. But in general there is no firmly established criterion. [164]

But, for our purposes, we really do not need to worry about these subtleties. They do not affect anything that we are concerned with, and almost never have any practical effect. Next, what is a Turing machine? The idea behind a Turing machine is to describe an extremely simple mathematical model of computation that is still capable of computing any function or implementing any algorithm. A Turing machine consists of an infinite tape divided into memory cells and a "head". The "head" moves over the tape and can read symbols from the tape and write symbols to the tape, one memory cell at a time. To picture this, think of a long horizontal tape on which there are vertical lines spaced an equal distance apart. Two consecutive vertical lines define a cell, and on each cell the head can either read a symbol or write a symbol. The set of symbols that the machine works with is called its "alphabet" and that "alphabet" is always finite. Additionally, at any given point in time, the machine is in one of a finite set of possible states. The operation of the machine is as follows. The head reads the symbol in the cell that it is currently over. Then, *based both on the symbol that it just read, and the state that the machine is currently in,* the head writes a replacement symbol into the *same* cell it was reading from, and then moves one step to the left or right, or halts the computation. The choice of which replacement symbol to write and which direction to move depends upon a *finite* set of choices: a table that specifies what to do for every possible combination of the current state and the symbol that was read. So, in reality, there isn't *one* Turing machine — they are really a class of possible computing devices. Each combination of the number of possible symbols and the number of possible states defines another possible computing device. The natural question then is: what is the most minimal Turing machine, i.e., one that has the fewest possible states and the smallest alphabet, that is universal? In the 1950s and 1960s, it was shown that the 2,2 Turing machine — that is, a Turing machine with 2 states and 2 symbols — was *not universal.* Later on, Marvin Minsky (an extremely famous computer scientist) showed that a 7,4 Turing machine — 7 states, 4 symbols — *was universal.* [44] Stephen Wolfram, in his book, *A New Kind of Science,* hypothesized that there existed a 2,3 Turing Machine — 2 states, 3 symbols — that was universal. [161, p. 709] Subsequently he offered a $25,000 prize for proving that it was universal and on October 24, 2007, it was announced that Alex Smith (then an undergraduate in Birmingham, England) had proved its universality.[7] Therefore, we now know that there exists a *universal* computer (i.e., a set of rules capable of computing any function or implementing any algorithm) that has just 2 states and 3 symbols! It is the smallest *universal* Turing machine that exists! It is instructive to look at the rules for this Turing machine. Figure 3.1 shows the rules for this: the most minimal *universal* Turing machine.

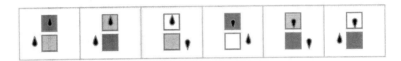

Figure 3.1: Rules for evolving the "2 State, 3 Symbol (color)" Turing Machine

There are 3 colors (white, light gray and dark gray) representing the "alphabet" of the machine. The black object in each row is the "head". The orientation of the black object—thin side up (figure 3.2), or thin side down (figure 3.3)—encodes the "state" of the machine and has exactly two orientations, up or down, corresponding to the two possible states of the machine.

Figure 3.2: Turing Machine Head "Up"

Figure 3.3: Turing Machine Head "Down"

Each rule shows at the bottom the successor state to the one on top. So, for example, if the machine was in state "up" and the cell below the head that encodes the symbol was dark gray, then the next symbol would be light gray, the head would be up, and it would have moved one cell to the left. Or if the current state was head down and the current symbol was light gray, then the next state would be head down, the cell would have turned to dark gray, and the head would have moved right.

Figure 3.4 shows what happens when we "run" this machine starting from the most basic configuration on the top row: all white cells and the initial state of the head being up.

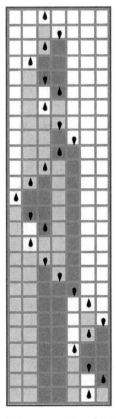

Figure 3.4: One possible execution history for the "2 State, 3 Symbol (color)" Turing Machine

3.5 Computational Equivalence

In fact, even more can be said about the phenomenon of universality. Stephen Wolfram originally put forth a proposition that he called the "Principle of Computational Equivalence". The principle states that *almost all processes that are not obviously simple are of equivalent sophistication.* So the operation of a computer, the set of states of a Turing machine, the output of many simple cellular automata, and so on, are all equally capable of doing sophisticated computing, and their sophistication is equal—they are all "universal"—capable of any possible computation: *Very simple sets of rules can be "universal": that is, capable of doing any computation!*

Figure 3.5 is a very simple example and is one of the simplest examples of a set of rules that is universal. There are white cells (0s) and black cells (1s). The rules show how to evolve cells by applying the rules. The top row specifies the state of a cell and its two nearest neighbors to the left and right. The bottom row says what the new cell below is if the cell on top, and its two nearest neighbors, match the given pattern. Under Wolfram's cellular automaton numbering scheme, this automaton is numbered 110, hence this is the "Rule 110 cellular automaton."[8]

Figure 3.5: Rules for evolving the "Rule 110" Cellular Automaton

Now, let us see what happens when we evolve this cellular automaton. Figure 3.6 shows the evolution of the Rule 110 automaton for 30 steps, starting from 30 randomly chosen black and white cells.

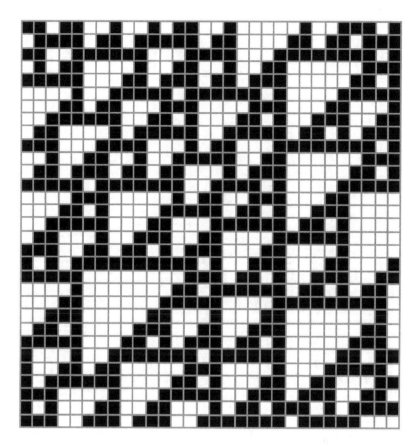

Figure 3.6: The "Rule 110" Cellular Automaton evolved for 30 steps starting from 30 randomly chosen black & white cells

We can see some repetition, and some structure, but none of it is obviously simple and none of it is obviously periodic. We can try running it from a very simple initial condition and seeing if that changes anything. Figure 3.7 evolves Rule 110 for 30 steps starting from just a single black cell and 30 white cells.

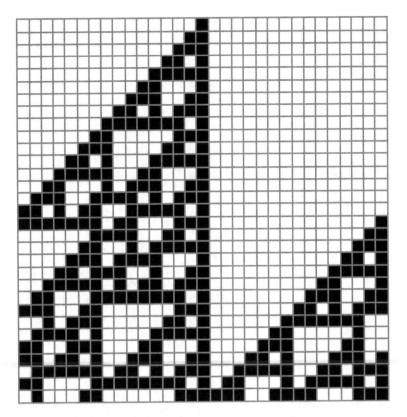

Figure 3.7: The "Rule 110" Cellular Automaton evolved for
30 steps starting from 1 black cell and 30 white cells

Again, we see no obvious simple structure, and no obvious periodicity. So we can *assume*, via the principle of computational equivalence, that this set of rules is universal. And, as it turns out, this automaton *is* universal! [34]

Let's just pause and reflect on that for a moment. A few rules, having to do with a collection of black and white squares, turns out to be a system we can use for any possible computation! This very strongly suggests that universality is not something special — something that only occurs rarely or has to be very carefully set up via clever thinking and engineering — but rather, it is a ubiquitous property of the world we live in. *Universality is everywhere.*

To be clear, many rules produce simple, periodic behavior that is *not universal* and it is completely obvious from looking at them that they're not universal. As an illustration of this, Figure 3.8 shows the results for running rules 100 through 139 starting from a single black cell [161, p. 54]. It is immediately apparent that rule 110 has behavior that is not simple, and that essentially all the other rules produce very simple behavior. Wolfram's conjecture is that rules that aren't universal (i.e., essentially all of them but rule 110 in this figure) are immediately apparent simply by looking at them.

An even simpler example of a universal computing system is OISC (One Instruction Set Computer): a *universal* computer based on precisely *one* instruction. There are OISCs based on manipulating one bit at a time (FlipJump); there are OISCs based on copying one bit and then jumping to a new instruction at a different memory address (BitBitJump); there are OISCs based on arithmetic (subtract and branch if less than or equal to one, subtract and branch if positive, subtract if positive else branch, etc.). And that's all, folks! *One instruction* suffices for universal computing! All possible computations can be constructed from any one of these examples! And that means something very profound, which is worth repeating: *"All of computing, and therefore every single thing in the universe that is computable, may be computed from a repeated combination of a single operation!"* Now it may be arbitrarily hard to figure out how to encode a computation we want done into a form that works for the given set of rules. And it can also be arbitrarily hard to figure out how to translate between these sets of rules (and, in fact, when proving that such and such set of rules is universal, the hard part is to translate those rules to another set that is known to be universal). This is why, in practice, for a given field of study, we create "blobs" of knowledge (i.e., theorems) that we can then put together to move further, rather than start from first principles every time. In economics, we start with supply

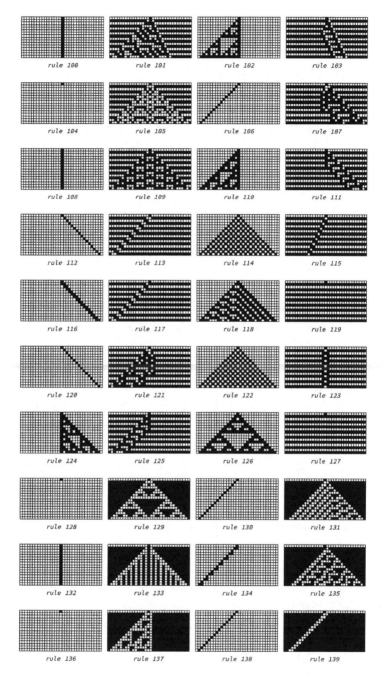

Figure 3.8: Rules 100 to 139 evolved starting from one black cell

and demand, each of which is a gigantic "blob" of knowledge consisting of gigantic bits of elementary "stuff". Similarly, in biology, we start from "evolution", in mathematics we (usually) start from known and already proved theorems, in computer programming we (almost always) start with a programming language not 1s and 0s, in astronomy we usually start with stars & galaxies (not electrons, protons & neutrons), and so on.

This leads to an interesting conundrum. If all these sets of rules are universal, then what's the difference between a rock and a human given that are both running on universal substrates? Because, in principle, the substrate that the rock is running on (electrons, protons, neutrons, etc.) is precisely the same as a human (also electrons, protons, neutrons, etc.). So what is different? The real difference is that each of them is doing a different computation *even though both are, in principle, capable of doing the same computations*. And for the computations that they do do, they are each the most efficient forms of that particular computation. But it gets very interesting here! The difference between a rock and a human is that a rock's behavior is almost perfectly predictable whereas a human's is not. The rock does not have free will in a precise proportion to how predictable its behavior is. And a human (or a dog, cat, wolf, sheep, goat, snake, beetle, etc.) has free will precisely because we cannot predict what anyone else is going to do: in fact, we cannot even predict what *we* are going to do! We therefore end up with a hierarchy of quantities of free will. Bacteria, for example, are unlikely to have any free will at all (or at most a very tiny bit) because they are essentially sense and respond creatures whose entire life is made up of direct reactions to stimuli, and so their behavior is highly predictable. And from there we can work up the chain of decreasing predictability: single celled creatures, multicellular creatures, more complicated animals consisting of differentiated and specialized organs such as reptiles, mammals, amphibians, fish, cetaceans, arthropods, etc. Human free will, in essentially every important sense, consists of humans choosing to do things differently than their "built in" programming.

3.6 What Is Science?

As we have seen, even simple rules can produce arbitrarily complex output which cannot be predicted except by just allowing the computation to happen: i.e., no shortcut to the result exists (i.e., computational "irreducibility"). In that case, what is science? And, even more generally, what is economics, law, sociology, anthropology, chemistry, or any other discipline? Because, if the output of even simple rules is unpredictable, then there are no regularities for us to base any field of "knowledge"! Interestingly, the answer to this conundrum is that arbitrarily complex output contains within it "pockets" of reducibility and all human knowledge consists of finding such pockets and understanding them. In fact, if you ran a computationally irreducible computation for an infinitely long time, you would find an infinite number of such pockets.

This means that humans (and AIs) will always have work to do (really, an infinite amount)—it shows, once and for all, that the "lump of labor" fallacy[9] that non-economically trained people tend to get caught up in is entirely false, *because it has to be!* It also means that humans (and AIs) will always have prejudices and biases. And not just for the tautological reason that, everything is deterministic, biases and prejudices exist, and therefore they are a necessary outcome of a deterministic paradigm. That's true, but also essentially a useless statement because it doesn't explain anything. In fact, and more interestingly, prejudices and biases are actually a *necessary outcome of computational irreducibility.* Computational irreducibility, by definition, means that there is only a limited amount of "stuff" that we can understand—that which is a pocket of reducibility. But furthermore, it *also* means that there is only a limited amount that we can ever understand which is a portion of a portion. We can only understand that portion of the universe which is *reducible*, and of that portion, only that which we can hold in our minds! Because all we can ever perceive is a portion of the infinite number of reducible portions of the universe, we can only ever perceive *some* of the whole. *Precisely because we can only perceive some of that whole, we must have biases and prejudices, because it is not possible for us to perceive everything.*

3.7 Organizational Free Will

What about organizations? Economics teaches us about the "fallacy of composition" [147]. For example, in voting, we have the "Condorcet paradox": essentially that even if every voter has rational preferences, the collective choice that is made by majority rule is not always rational. For example, when you have groups of voters, it is extremely easy to have a situation where majority of voters prefer A to B, prefer B to C, and yet prefer C to A! That's because when you have collective choice the result is not "transitive." The problem is that the conflicting majorities are made up of different groups of people. [144] Let's take a particularly simple example to see how this might happen. Suppose that we have 3 candidates, A, B and C. And there are three voters, $V1$, $V2$ and $V3$. Let's say that the voters have the following preferences. (Read > to mean prefers so $A > B$ means A is preferred to B.)

- $V1 : A > B > C$
- $V2 : B > C > A$
- $V3 : C > A > B$

So, who "wins" this election? If you declare A the winner, then $V2$ and $V3$ would have preferred anyone else. If you declare B the winner, then $V1$ and $V3$ would have preferred anyone else. And if you declare C the winner, then $V1$ and $V2$ would have preferred someone else. *No matter who you declare as the "winner" of this election, most voters prefer someone else!*

Similarly, if you try to predict the outcome of other collective decisions from committees, courts, political parties, and so on, you will always have a prediction error. The preferences of the group may not be particularly predictable in practice: think polling error. For example, we need a Supreme Court *because* it is unpredictable. The very fact that we do not replace the Supreme Court with an algorithm tells us that we believe that the Supreme Court *as an institution* has free will. Therefore, under our predictability paradigm, *prediction error* is a form of "free will"! And, in fact, it is the *existence of that prediction error that makes it worthwhile to have collective judgments.* Because if there was no prediction error, there would be no need whatsoever for these collective outcomes—we could simply write an algorithm that would output the result and that would be that. One excellent way to think about all this is via the concept of Computational Irreducibility. The more reducible something is, the more predictable it is since you can shortcut the result without having to go through the whole process.

3.8 Minimality

A fun consideration is to think about how "minimal" a brain can be and still have interesting biological behavior like learning, transmission of learning between individuals, emotions, and other so-called "higher" functions. We are used to the idea that large mammals share quite a bit of intelligence with us, so I don't think most people would find it particularly surprising that pigs reconcile after fighting, pig bystanders offer consolation to pig losers, and other pigs can step in as peacemakers. [117] Or that raccoons are able to open a transparent box holding a snack or be able to figure out that if a latch on a door couldn't be opened, they had to slide open a window. [69] Well, how about smaller critters? For starters, rats can imagine places and objects they can't see, have rich inner lives [116], and while sleeping practice running mazes, cats rehearse hunts, and zebra finches practice their songs. Even zebrafish sleep and dream. "We are not the only species capable of remembering and learning"! [107] Invertebrates such as cephalopods (for example, octopus, squid, cuttlefish) are able to process information quickly so that they can change their shape, color, and texture to blend in with their surroundings. They communicate with each other, are capable of spatial learning, use tools to solve problems and can *even get bored!* [136]

But wait, there's more! Much more! Crows, for example, are able to learn recursive skills, previously thought to be unique to humans.

A facet of complex thought long believed to be unique to humans may be for the birds, too.

Crows can learn a skill previously believed to distinguish the minds and communication of humans from other animals, according to a study published on Wednesday in the journal *Science Advances*. The research could help bolster evidence that recursion, defined in the study as the cognitive ability to embed a structure within similar structures, is a skill that some animals, like some monkeys and birds, can learn.

Two crows appeared to grasp the cognitive concept at a toddler's skill level, the study said. The crows required just a few days of training before they learned the skill. [101]

And,

Trained crows outperformed monkeys and matched human toddlers in identifying paired elements buried within larger sequences, according to a study published in *Science Advances*. The trait, known as recursion, is considered a basic building block of language. Crows have also been found to use tools, understand the concept of zero and follow basic analogies. [41]

Crows can make friends, play hide and seek, recognize faces, are aware of their surroundings, have families, defend their spaces, engage with other crow communities [111], hold funerals [72], and like primates understand probability. [120]

Going smaller, brainless box jellyfish can learn [62], and jumping spiders "rapidly move their eyes and twitch during rest, suggesting they have visual dreams, never before observed in arachnids." [21] Or, how about wasps and bees both of which have incredibly tiny brains? And yet:

Paper wasps can learn to distinguish between pairs of stimuli that are the same or different, demonstrating a grasp of abstract concepts. Such cognitive abilities have only been shown in a relatively small group of animals, including some birds, dolphins and honeybees. [140, 160]

"We now have suggestive evidence that there is some level of conscious awareness in bees—that there is a sentience, that they have emotion-like states," says Lars Chittka, professor of sensory and behavioural ecology at Queen Mary University of London. [...] "[T]hey can count, recognise images of human faces and learn simple tool use and abstract concepts." [55]

Prof. Chittka's experiments suggest that bees recognize human faces, count landmarks, identify a sphere visually which previously they could only feel in the dark, and understand concepts like "same" and "different". Amazingly, he has also shown that bees like to *play* with balls: "playfulness is not restricted to mammals and birds"! [22] Bees can also learn from each other and improve on each other's techniques without trial and error. They have emotions: for example, an experiment induced them to have a kind of Post-Traumatic Stress Disorder. Bees can learn numbers higher than four if we train them the right way. [50]

Honeybees in particular are the only animals other than humans that can tell odd numbers from even numbers. [71] They also show "withdrawal symptoms when weaned off alcohol" [169] and they "use a 'mental number line' to keep track of things"! [170] In other words, bees are sentient creatures that have feelings! And yet, the brain of a bee is two cubic millimeters in size, or 0.0002% of the human brain! [48] In that incredibly tiny space, bees have the capability of navigation, emotions, thinking, counting, learning, memory, etc. Looked at from the point of view of Computational Equivalence, this is no surprise. It doesn't take very much "computing power" (in terms of the number and kinds of rules that need to exist) to get a universal computer. Is the bee's brain a universal computer? We don't know, yet, but it is very likely to be universal.

And this is why we constantly get surprises in biology. We cannot even predict what sorts of mathematical creatures may live in a given cellular automaton with very simple rules. The only thing we can do is to run the automaton and observe what is obtained. But we have no way of predicting what is going to come. Similarly, all sorts of things that biologists thought "were so" turn out not to be quite so hard and fast. For example, biologists are finding that evolution may not in fact form a "tree of life". There is a bird called a hoatzin, that lives in South America. It doesn't seem to fit in any sensible way into the tree of life as we currently know it.

DNA research has not solved the mysteries of the hoatzin; it has deepened them. One 2014 analysis suggested that the bird's closest living relatives are cranes and shorebirds such as gulls and plovers. Another, in 2020, concluded that this clumsy flier is a sister species to a group that includes tiny, hovering hummingbirds and high-speed swifts. "Frankly, there is no one in the world who knows what hoatzins are," Cracraft, who is now a member of B10K, said. The hoatzin may be more than a missing piece of the evolutionary puzzle. It may be a sphinx with a riddle that many biologists are reluctant to consider: What if the pattern of evolution is not actually a tree? [40]

From the point of view of computational equivalence, this is no mystery. Biology is too complex to admit large quantities of detailed reducibility. And without those pockets of reducibility, all we can do is observe and classify—we cannot predict. The eminent physicist Ernest Rutherford was reputed to have divided science into two categories: physics and stamp collecting! The real reason for this division is physics works with phenomena at their most basic level, where large quantities of reducibility exist, which means it can make hard predictions. Every other field, from economics to biology, from anthropology to sociology, from political science to international affairs, has to deal with phenomena that contain very little reducibility, and that reducibility which they do contain is at best very approximate.

3.9 Universality: Consequences

Another interesting question is about universality. If two things are universal, then there exists an arbitrarily complicated encoding that allows us to express one thing in terms of the other thing. So why can't a dog compute a human and vice versa? Both are running on a "universal" substrate: the laws of physics. Remember that the difference between a dog and a human is *only in the arrangement of their atoms*. The issue is that it will be arbitrarily complicated to rearrange the atoms that make up a dog into atoms that make up a human and vice versa even though both sets of atoms are identical in every respect! (Yes, the amounts of some atoms may be different in a dog than in a human, but that's easily solved in a thought experiment by using energy to transmute the atoms as needed. This is a direct consequence of the famous $E = mc^2$.)

In fact, a simple set of facts is all that is needed to convince yourself that it is the arrangement, and only the arrangement, that differentiates one human from another, or a human from a dog, etc. The human body renews itself at different paces. Red blood cells last about four months, skin cells two to four weeks, hair from three (women) to six (men) years, liver cells 150 to 500 days, intestinal cells five days, skeletal cells ten years, and so on [114]. If those cells are being replaced then, by definition, the precise cells that make up a human are not what is important or relevant. And if the precise cells don't matter, then what really matters is the *content of those cells*: in other words, *their design*. For example, some people have better oxygenation capabilities than others—an adaptation to living at high altitudes.

Highland populations in Tibet, such as the famous Sherpas who serve as Himalaya Mountain guides [...], have lived at high altitudes for only about three thousand years. Their adaptations to high altitude include an increase in the rate of breathing even at rest without alkalosis occurring, and an expansion in the width of the blood vessels (both capillaries and arteries) that carry oxygenated blood to the cells. These changes allow them to carry more oxygen to their muscles and have a higher capacity for exercise at high altitude. Their adaptations to high altitude occurred very rapidly in evolutionary terms and are considered to be the most rapid process of phenotypically observable evolution in humans [1].

In other words, their cells changed because their DNA adapted to the environment they lived in. But what does it mean that their cells changed? After all, their cells are changing all the time, according to the schedule laid out above. What it means is that the information encoded in their DNA changed, which changed the cells that make up their body in a specific way. And when you change the cells that make up someone's body, what you are changing is the precise arrangement of the molecules that make up those cells: it is the *arrangement, and only the arrangement*, of those molecules (that are made up of atoms that are made up of particles that are the excited states of quantum fields) that matters. And, in fact, it gets even more interesting. Suppose we have some technology from the future: let's say, the technology from *The Empire Strikes Back*. Luke Skywalker loses his right hand in a duel with Darth Vader. At the end of the movie, he is given a fully-functional, bionic, right hand that looks to the naked eye exactly like a regular hand. There is, obviously, no law of physics that forbids us from creating this technology, which means one day it might exist. So this raises the question: if we replace Skywalker's right hand with a new, bionic, hand, is it still Luke Skywalker? I suspect everyone reading this will agree that it is. Well now suppose our technology is even more amazing, and we replace not only his hands, but also his legs. Is it still Luke? What if we now replace his heart with an artificial one? And then the spleen, pancreas, kidneys, and liver. Is it still him? And so on: we could imagine a technology, none of which is forbidden by the laws of physics, that replaces every single atom in his body with atoms of our creation. Is it still Luke? Or consider the Japanese Ise Jingu grand shrine in Mie Prefecture.

Every 20 years, locals tear down the Ise Jingu grand shrine in Mie Prefecture, Japan, only to rebuild it anew. They have been doing this for around 1,300 years. Some records indicate the Shinto shrine is up to 2,000-years old. The process of rebuilding the wooden structure every couple decades helped to preserve the original architect's design against the otherwise eroding effects of time. "It's[sic] secret isn't heroic engineering or structural overkill, but rather cultural continuity," writes the Long Now Foundation. [109]

Has this been the same shrine over the last 1300 years, or have there been 1300/20 = 65 shrines? Everyone involved considers it to be the same shrine. The answer is clear: if the patterns that represent Luke or the shrine are present in the new replacements, then yes, it is still Luke or the shrine. If those patterns change, then it isn't Luke or the shrine anymore.

And there is good evidence for this assertion as well. Consider the case of Phineas Gage. In 1848, an iron rod went through Gage's left cheek and into his brain, destroying his frontal lobe along the way. After his accident, "his personality and behavior were so changed as a result of the frontal lobe damage that many of his friends described him as an almost different person entirely." [28] Legally, he was the same person: as far as the law was (and is) concerned, nothing had changed. However, because the patterns encoded in the molecules of the frontal lobe are critical to a human's personality, the fact that those patterns were disrupted meant that this was a *new* Phineas Gage, and *no longer the old one!*

The evidence that everything in your brain is all physical is incredibly strong. Those that want to argue against this are essentially engaged in wishful thinking. An example of this physicalism is depression. It has become increasingly clear that it is a disorder of how the brain is functioning. A patient called John volunteered to have electrodes implanted in his brain to alleviate his depression. What happened next is quite amazing:

[t]o find the right place to stimulate, the team needed to wake John up during his operation. He remembers being repeatedly asked how he felt as surgeons probed his brain with electrodes. "Then they hit a spot and I said: 'I actually feel back online,'" he says. "Depression is like a constant weight on your soul. *When they touched that perfect little spot, that weight lifted.*" [emphasis added] [65]

By the way, universality is interesting for another reason. As I said earlier, in section *2.23, A Paradox,* I find it a paradox to contemplate how a part of the universe could understand the whole. From physics, we get Max Tegmark's argument that it is simpler to describe the whole rather than a part of the whole which resolves the paradox. But universality gives us another resolution. The universe contains within it universal computers because we already know they exist: we have built them and use them every day. And now, as per section *3.8, Minimality,* we know that universal computers can be incredibly minimal. So a minimal computer that is part of the universe can simulate the entire universe because it is equivalent in its computational sophistication to the whole universe! This is another reason why it is possible to write a model of the universe from within the universe!

3.10 Death

If the only thing that differentiates one human from another or a human from a tree or a tree from a dog is the arrangement of atoms, then what is death? Death is, obviously, the disruption of that arrangement to the point where that arrangement (at least with current technology) can no longer be restored. As Nick Lane puts it:

The difference between being alive or dead lies in energy flow, in the ability of cells to continually regenerate themselves from simpler building blocks. [...] Metabolism is what keeps us alive—it is what being alive is—the sum of the continuous transformations of small molecules on a timescale of nanoseconds, nanosecond after nanosecond. If we live to the age of eighty, we will have lived through nearly three billion billion (3×10^{18}) nanoseconds-worth of metabolism. No wonder we run down. [86, p. 4]

This tells us several things. One, death is *not* inevitable. There is no law of physics that says that you cannot restore the arrangement of the atoms that make up a human or, for that matter, any other creature. As long as you have the technology and the energy to do it, you can indefinitely restore any arrangement of atoms. Two, it tells us that the loss of that arrangement is a loss of information. Whenever any creature dies, whether it be an ant, a butterfly, a whale, or a human, some of the information that was contained in the arrangement of the atoms that made up that creature is lost. Third, it tells you that, *if* you had some way of recording that information, *then at least in theory*, you would be able to "back up" that creature so that you could restore it (or make a copy). Fourth, that what is unique about any given creature is *the information encoded* in that creature, which includes both the information the creature has at birth, *and* the information it has acquired since birth. So when your parent dies, for example, what you miss is the ability to be able to interact with them—the information encoded within you "misses" being able to interact with the information encoded in your parent.

3.11 Networking

The Internet runs on a protocol called TCP/IP (Transmission Control Protocol/Internet Protocol). TCP is the protocol that takes a stream of octets (an octet is 8 bits = 1 byte) that originates in one application running on one computer and delivers them in a reliable, ordered, and error-checked way to another application running on another computer. That other computer is physically somewhere else, whether nearby or far away. IP runs on top of TCP and its job is to take "data packets" from the source to the destination based solely on the "IP address" contained in the header of the packet. To do this, IP defines a "packet structure" that encapsulates the data to be delivered (the data being nothing more than a string of 0s and 1s) and it defines a way of addressing that packet, so that the packet knows where it comes from, and where it is going. Internet packets are sent ("routed") from one "host" (i.e., a computer on the Internet) to another "host" until the packet gets to its destination: it is a point to point protocol (like Southwest Airlines). So if you wanted to send a packet across the world, it would bounce from host to host until it got to its destination. Each host has a "routing table" that contains information about where to forward a given packet so that it gets to its destination in a reasonably efficient way. Sometimes, however, a "packet" can get lost. This can happen for a number of reasons: hardware issues, bugs in software, network congestion and so on. The key point is: a computer is only as useful as the information it contains and can communicate to another computer. If the computer finds itself entirely unable to communicate, it might as well be dead. And this is where there is a strong analogy between a network and a human (or any other living creature). If there is too much packet loss in a network, then the hosts connected to it become progressively more and more useless—i.e., dead. Similarly, if a human is entirely unable to communicate, the human is effectively dead: that human has 100% packet loss. This is why "locked-in" syndrome—where a person is alive but entirely unable to communicate—is so terrible for them. For all intents and purposes, a human is essentially a machine to communicate. Locked-in syndrome altogether removes the ability for a human to accomplish this function. Physical death does the same thing—it permanently removes the ability for a specific human to communicate: in other words, permanent, 100% packet loss. Therefore, *each human forms a node on a network of communication that encompasses all living beings on the planet. We are part of the environment in the most fundamental possible way: disconnected from our network, we would no longer be human!*[10]

3.12 Predetermination

Just because something is *predetermined does not mean that it is predictable.* The main reason people have trouble with accepting the deterministic paradigm is that it means that everything is inevitable and no choices have to be made. But again, this is an incorrect framing of the problem. The real issue, the real discomfort, is with inevitability (i.e., predictability) not determinism, and conflating the two is an error. However, for every possible practical purpose: *Who cares if something is in some sense inevitable if it is entirely and always unpredictable?* It is hard to grasp this intuitively because in normal life, if something is inevitable, it is also predictable. But here we have something, i.e., free will in practice, that is entirely unpredictable and yet entirely inevitable. And that is what causes the confusion.

3.13 Fate

Predetermination can be thought of as "fate". Inescapable fate is an old and recurring idea. For example, in "Germanic mythology, gods were not the highest powers. The highest power was fate." [128] Usually fate is thought of as something that a supernatural power determines. In our case, there is no necessity for a supernatural power (although we could have one compatible with determinism: see section *2.16, Compatibility with a Deity*), and fate comes from the inexorable march of the laws of physics. Fate really exists, and really is inescapable. *Fate exists because it must!*

Somehow this seems surprising to people. But it shouldn't. We are made up of the same stuff as the physical world. So we will follow the same laws as the physical world. For example, we can use the theory of relativity, and a bunch of observations, to make a prediction about the future fate of the universe as a whole. Depending on how much matter there is in the universe, and how dense that matter is, we can calculate if the universe will expand forever, or if it will at some point stop expanding. Using Astrophysics, we know that the sun will, in about five billion years, expand and turn into a red giant star, engulf Mercury and Venus, and possibly engulf Earth itself. When the sun does become a red giant, life on earth will be impossible. No serious scientist doubts that this will happen. That is the ultimate fate of all life on Earth and is an inescapable fate for the Earth and any inhabitants left upon it at the time. However, the reason we can make this prediction with great certainty is that astrophysics (and all of science) is the art of discovering pockets of computational reducibility: pockets where we can discover laws to shortcut having to run the whole system. All of science is the art of finding these shortcuts. These shortcuts allow us to predict the future rather than having to wait to see it unfold to know what happened. And that is the only difference between our fate as humans, and the fate of an object like the Sun. The physics of the Sun that we are interested in allows massive amounts of computational reducibility so we can calculate what will happen to it. Our personal physics, essentially because we are so very much smaller than the Sun and care much more about the behavior of our constituent parts (and are therefore much more complex to model), has few or no pockets of reducibility and so we cannot predict our own fate. That is the difference between us and an object in the physical world about which we can make predictions: *complexity*. It is that complexity, *and only that complexity*, that makes it seem like we have Free Will in Theory, even though we do not and *cannot*.

It is more or less a canonical rule that big things are easier to predict than little things. One reason is the one mentioned in section 2.23, *A Paradox*—that the description of a part of a whole can be arbitrarily more complex than a description of the whole. But there are other reasons too. The law of large numbers, for instance: the more often you perform a given action that has some randomness (i.e., lack of predictability) in its outcome, the closer the *average* of the results is to the expected value of that action. A given spin of a roulette wheel can produce any of the outcomes on the wheel, but over the long run, the outcomes will all even out and all that will be left is the house's edge (which is how casinos make money—their expected value is positive—so why anyone gambles remains a mystery to me since all those gambles have a negative expected value). Another reason is that with big things, we are typically interested in some "averaged" behavior— we almost never care about the behavior of its constituent parts. Similarly, if you want to know how a gigantic ball of gas behaves (such as the Sun), you can forget about the movement of any individual molecule in that gas, and just model the whole. Any unknowns in the movement of individual gas molecules are averaged out over the gigantic collection of molecules that are of interest. You can't precisely predict the height any given child will reach (you can get an idea by looking at its parents, but merely an idea—there is still plenty of variation), but you can precisely model the outcome of the heights of all the people on Earth: they will follow what is known as a "Normal" or "Gaussian" distribution. This predictability shows up in many cases and those cases are ones where computational reducibility rules. It isn't always true, of course. Why, for example, is the sun predictable but the economy isn't?

> [M]icroeconomists are people who are wrong about specific things, and macroeconomists are wrong about things in general. [15]

The problem for economists is that they're dealing with an incredibly complex system, in essentially every case they consider. The quantity of computational reducibility is extremely small, despite the fact that the systems they are studying are in many cases very large. The complexity comes from the fact that there are a large number of different constituent entities, and those entities are interacting with one another. As a result, averages are unstable and the law of large numbers is not particularly helpful. Many consistent principles do stand out: if you raise the cost of something, less of it will be demanded;[11] if you tax something more, you will get less of it; circulating large quantities of extra cash in an economy will lead to inflation. And so on. But how much will happen? When will it happen? How long will it take to happen? These are all very difficult to answer as they are all contingent upon the current state of the system under consideration.

> COWEN: Now, during the pandemic, government debt is way up. Production is not up. What's the prediction of the fiscal theory about price inflation?
>
> COCHRANE: Economists should never make predictions.
>
> COWEN: But theories make predictions.
>
> COCHRANE: Theories make conditional predictions. Theories make if x happens and you hold everything else constant, then y ought to happen. Economics is awfully bad at making unconditional predictions. "Here's what's going to happen." Period. I think it's a mistake to get into that game, but let me tell you what fiscal theory says. [32]

So what this tells us is that if we have a system with a large number of constituent parts, we can model them as long as their interactions are limited. The more complex those interactions become, the harder it is to model the system. *Interactions amongst*

constituent parts reduce the quantity of computational reducibility.
So although the economy is produced by a completely
deterministic underlying system (i.e., the laws of physics), we
frequently find it very difficult to find much reducibility within
it to make good predictions.

3.14 Chaos

Chaos theory, i.e. the mathematics and physics of chaotic
behavior, is often thought of as random. But in fact, chaos is
entirely deterministic. What is special about chaos is that
prediction is *impossible* (except, and even then approximately
and only for short periods, in some special circumstances) while
the underlying system is *entirely* deterministic. Chaos theory
says that, even if we know the complete rules of a given system,
any uncertainty in the initial conditions of that system will lead
to exponentially different outcomes in a short period of time.
A famous example is that a butterfly beating its wings in Hong
Kong this month can affect a storm system in New York next
month. Why?

Because the equations for predicting the weather are
exquisitely sensitive to their inputs. Infinitesimal changes in
the inputs, well below any possible detection threshold, can
produce vastly different outputs.

Wikipedia has a beautiful illustration of this in a double-
pendulum system. [145] As is immediately apparent, while the
three double-pendulums in the figure start off at almost exactly
but not quite the same position, their subsequent motion is
entirely unrelated to each other! And yet, the double-pendulum
is an entirely deterministic system. If you are curious, Wikipedia
has a nice animated visualization. See [145].

Here are some figures to illustrate the point further.[12] Figure
3.9 is an example of a double-pendulum. The top ball is hanging
freely, and the bottom ball is hanging freely from the top ball.
Both balls are connected to each other, and to their hanging

point with springs. Each ball has a mass, an associated starting position, a spring length, and a spring constant. The latter measures how "springy" the spring is. Each pair of graphs shows the beginning state of the pendulum, and the state after 50 seconds have passed. The squiggly bottom lines are the trajectory that the bottom ball had during those 50 seconds, and the squiggly top lines show the same thing for the top balls.

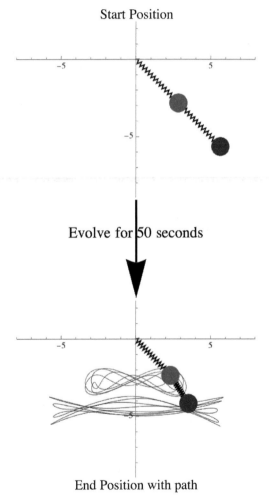

Figure 3.9: Initial Double Pendulum

Next, let's change the starting position a little bit. If we do that, we get figure 3.10.

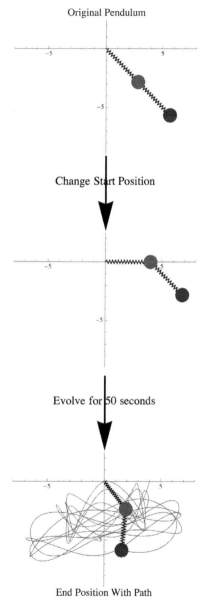

Figure 3.10: Double Pendulum – change in starting position

Similarly, we can change the mass of one of the balls to get figure 3.11.

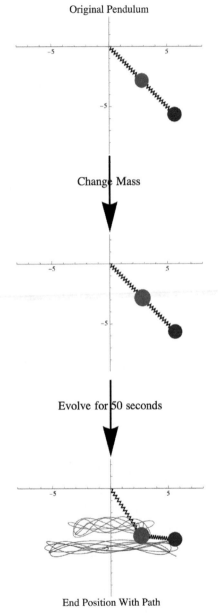

Figure 3.11: Double Pendulum – change in mass

We can also change the "spring constant" which signifies how "springy" the spring is. If we do that, we get figure 3.12.

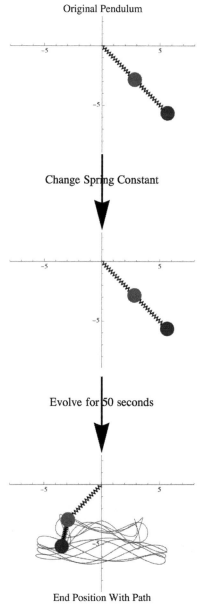

Figure 3.12: Double Pendulum – change in spring constant

And, lastly, for completeness, let's change the length of one of the springs which produces figure 3.13.

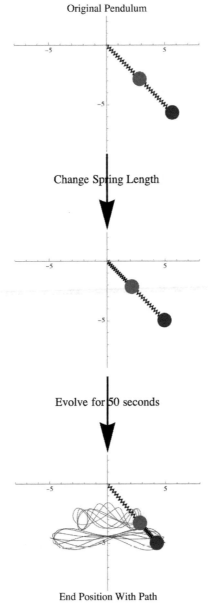

Figure 3.13: Double Pendulum – change in spring length

Note how making slight changes in any parameter—starting position, mass, spring length or spring constant—produces entirely different trajectories and entirely different outcomes. This is an example of chaos: "sensitive dependence on initial conditions". And yet, the system is entirely deterministic! So in this condition, prediction of the future path of the system would be impossible, because it is impossible to specify initial conditions *exactly* as they are real numbers! And you cannot encode real numbers exactly, in the same way that you cannot ever encode π exactly. So this is another limit on our ability to predict. Let's call this the "chaos limit to prediction".

3.15 Irreducibility versus Chaos

When one does "science", the usual goal is to find a mathematical formula that allows one to calculate the result of running a given experiment, without having to actually run it. Newton's famous laws, for example, allow us to calculate the velocity of an object dropped from some given height. We do not need to actually perform the experiment, as the formula will tell us the answer. So, for example, if an object is dropped at time $t = 0$, then (neglecting air resistance) at some other time, t, its position will be given by $v_0 * t - 1/2 * g * t^2$, where v_0 is the initial velocity, g is the acceleration due to gravity (the negative sign is because the acceleration is towards the earth, i.e., in the downward direction), and t is the time (the number of seconds that have elapsed since the object was dropped) at which we want the answer. This law is an example of "computational reducibility": we do not need to do the entire "computation" (i.e., dropping the object and measuring its velocity after time t) to know the answer. Science in general sets itself up to work largely on such problems: ones where there is reducibility. However, many problems are not reducible. And, most interestingly, those problems *do not need to be complicated*. Even very simple rules can produce output that is not reducible; i.e., it is irreducible.

For systems governed by such rules, you cannot find a formula that shortcuts the computation. You have to "do" the entire computation to find out the result. By the way, as a general rule, proving that some set of rules is computationally irreducible is very difficult since you are trying to prove a negative. The usual method is to see if there is a way to prove that the set of rules is "universal" because a universal set of rules must be irreducible (else they wouldn't be universal). One typically proves universality by showing that the set of rules in question is equivalent to another set of rules that is already known to be universal. In fact, it is the very presence of irreducibility that makes it hard to prove the presence of irreducibility, which is beautifully ironic!

Therefore, computational irreducibility goes a step further than chaos. *Even if* we have the rules of the system, and *even if* we have the *exact* initial conditions (i.e. inputs) to the system down to every relevant decimal place, *it may still take an irreducible amount of time to do the computation.* In other words, the amount of time taken to do the computation affords no shortcut: to see the result, you simply *have* to march through every step of the computation; there is no other possibility.

Here is a simple example. Figure 3.14 shows the rules for the "Rule 30" cellular automaton (compare its rules to the Rule 110 automaton in figure 3.5).

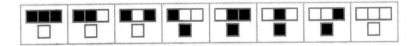

Figure 3.14: Rules for evolving the "Rule 30" Cellular
Automaton

Let us see what happens when we evolve this cellular automaton. Figure 3.15 shows the evolution of the Rule 30 automaton for 30 steps, starting from one black cell. As is immediately apparent, there is no simple behavior for this cellular automaton. Its behavior seems to be random and there seems to be no way to represent its evolution via a formula—in other words, it is *irreducible*.[13]

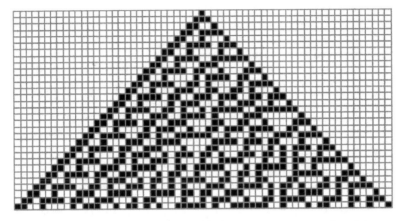

Figure 3.15: The "Rule 30" Cellular Automaton evolved for 30 steps starting from 1 black cell and 30 white cells

In general, therefore, there is no reason for us to expect that any given set of rules will produce results that can be "shortcut" via formulas, even in incredibly simple systems. And if that is the case, then irreducibility really is everywhere. And that means that, in general, unless a situation is chosen with extreme care (which is what we do in science experiments) making predictions will be impossible. *Even with perfect information we will still be unable to predict human behavior due to irreducibility.*

3.16 Information Loss
Physics postulates (except for the one exception of the Copenhagen interpretation of QM) that information is never lost. All of physics is built on this postulate (and seems to work quite nicely).[14]

FWIT means information loss. For a choice to be made, *in a nondeterministic manner*, doesn't just mean that the "choice" must run on a nondeterministic substrate. It also means, necessarily, that information must be lost in making that choice. This is because there is no way, even in principle, of reconstructing the information once a choice has been made. Therefore, FWIT is in contradiction to the postulate of the conservation of information.[15]

Philosophically, this is similar to the problematic definition of a quantum measurement in the Copenhagen interpretation of quantum mechanics. So, if we are to believe the Copenhagen interpretation, the collapse of the wave function is a "choice"![16]

3.17 The Game of Life

You cannot be serious!
—John McEnroe

Ah, but I am serious about using computation as a paradigm for thinking about determinism and free will. One reason is the incredibly strong set of hints given by an invention of John Conway's called *The Game of Life*. In essence, it is a cellular automaton, but with a very specific set of rules that encode birth, death, and reproduction. The game consists of a two-dimensional grid of cells, some of which are live (black) and some of which are dead (white). Each cell interacts with its eight neighbors: the four cells immediately above, below, right, and left, as well as the four cells diagonally above left, above right, below left and below right. You start with some initial pattern, which is the "seed" of the game. You then apply the following rules:

- Underpopulation: If a cell is live, but has fewer than two live neighbors, it dies.

- Overpopulation: If a cell is live, but has more than three live neighbors, it dies.
- Life: If a cell is live, with exactly two or three live neighbors, it lives to the next generation.
- Reproduction: If a cell is dead, and has exactly three live neighbors, it becomes live.

You "play" the game by repeatedly applying the rules. Each time you apply the rules, you get a new row of cells. Each application of rules is called a "tick" as in the tick of a clock. The result of the game is *completely determined* by the initial configuration chosen. What is fascinating is the truly incredible variety of what happens as you watch the game unfold. You get patterns called "still lifes" that maintain their shape as the ticks go by; "spaceships" that move themselves across the grid; "oscillators" that return to their initial shape after a finite number of ticks; as well as many others. There are oscillating patterns that oscillate every two ticks, some that oscillate every three ticks, and there are others oscillating every four ticks, eight ticks, fourteen ticks, fifteen ticks, and thirty ticks. Then there are patterns that evolve for very long periods before stabilizing called "Methuselahs". There are others that evolve for very long periods and then disappear. Then there are patterns called "guns" that shoot out other patterns called "gliders" (a form of spaceship, because it moves across the grid). There are patterns that lay blocks: as the pattern moves, it leaves behind blocks; patterns called "puffer trains" that leave behind debris as they move; and patterns called "rakes" which emit spaceships as they move. The patterns interact with each other and can be used to construct a counter, a memory, and logic gates such as AND, OR and NOT. In fact, you can build a "universal Turing Machine" in the Game of Life, which means that the Game of Life is universal—it can be used to compute any possible computation. Tetris, for instance, has been implemented in the Game of Life!

There are self-replicating patterns of various kinds. For instance, there is a pattern called "Gemini" that makes a copy of itself while destroying its parent and takes about 34 million ticks to do it! [87] There is a pattern that can build a complete copy of itself, even including its own instruction tape. [49] And so on and so forth. The point is, these very simple rules are already showing us behavior of great complexity, and in fact, since these simple rules are "universal" they can show us arbitrarily complicated behavior since they are theoretically able to do any possible computation. Additionally, the Game of Life is undecidable. If you are given two patterns, you cannot tell if the second pattern must follow from the first one—no algorithm exists to let you figure that out. The only thing you can do is to let the game run and see if the second pattern occurs, and you may have to wait an infinitely long amount of time to know whether it does or it doesn't. And so, yet again, we see very simple deterministic rules producing behavior that is *undecidable, extremely complex, and lifelike*. The Game of Life is possibly the strongest hint that computation is the right paradigm for thinking about determinism, and thus free will in practice.

3.18 Traffic! Traffic! Traffic!

Despite traffic being one of the most infuriating things in the known universe, modeling it can be quite fun. Now that we have looked at different kinds of cellular automata, we can try to use one as a toy model to understand real world phenomena. Interestingly, the very simple "Rule 184" cellular automaton nicely models the main features of traffic and helps us start developing our intuition. The rules for this automaton are in figure 3.16.

Figure 3.16: Rules for Traffic: evolving the "Rule 184" Cellular Automaton

Think of traffic as moving from left to right and let us simulate one lane of traffic. Time will move down the page so each row is a snapshot of traffic one "tick" of time after the previous row. A black cell is occupied by a vehicle and a white cell is free. The rules say that a vehicle moves at a constant speed if there is an open space in front of it, else it stops. Figure 3.17 shows what happens to traffic when a quarter of the road is occupied by vehicles.

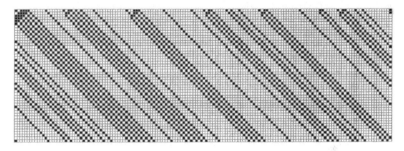

Figure 3.17: Quarter of the road occupied by vehicles

Next, figure 3.18 shows what happens to traffic when half of the road is occupied by vehicles.

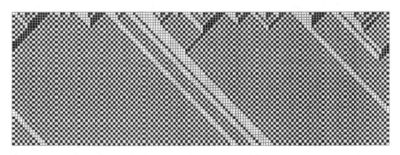

Figure 3.18: Half of the road occupied by vehicles

And then finally, figure 3.19 shows what happens to traffic when three-quarters of the road is occupied by vehicles.

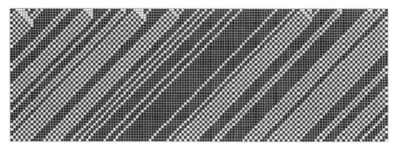

Figure 3.19: Three-quarters of the road occupied by vehicles

This is an *extremely* primitive model of traffic and yet it shows you traffic's essential features. When traffic is light, there are clusters of cars moving freely that are separated by stretches of open road. When traffic is heavy, you get *waves* of stop and go traffic, and those waves move backwards in the opposite direction to the flow of traffic. And so now we can see why traffic is so infuriating! As soon as traffic gets heavy, even if you have plenty of "free space" on the road, you get stop and go traffic *literally* for no emotionally satisfying reason at all!

Notes

1. Rolf Landauer figured out, back in 1961, that you generate heat whenever you lose information. Landauer's limit is the fundamental limit that applies to classical computers — so, for example, there is a limit to how efficient a silicon chip can ever be! There are attempts to create "reversible computing" such that information is never lost but those have not yet been successful.

2. In quantum field theory, if you reverse time, you also have to reverse two other quantities called Charge and Parity, so it is technically CPT invariance — but that doesn't change

the argument. Also, even if you ignore the "C" (charge) and the "P" (parity) and just time reverse the system, you can certainly get a bit of time reversal invariance violation. But, *even then* every successor state has a unique predecessor state, so *no information is lost* and the system remains reversible for our purposes.

3. See section *3.15, Irreducibility versus Chaos*.
4. See [94] for an example of an actual, working, OISC.
5. For a very nice introduction, see reference [162].
6. For example, reference [82].
7. https://www.wolframscience.com/prizes/tm23/.
8. See [17] for a very nice discussion of cellular automata, Daniel Dennett, Stephen Wolfram, and Free Will.
9. The lump of labor fallacy is the mistaken belief that there is a fixed amount of work available in the economy, and that increasing the number of workers decreases the amount of work available for everyone else, or vice versa. The fallacy begins with the faulty assumption that an economy can only support so many jobs—i.e. a fixed lump of labor. It is then applied to policy issues such as immigration: allowing more immigrants decreases jobs available for native workers. Economists regard this reasoning as fallacious because many factors impact required labor levels in an economy. For example, increasing the employment of labor can expand the overall size of the economy, leading to further job creation. In contrast, reducing the amount of labor employed would decrease economic activity, thus further decreasing the demand for labor. The lump of labor fallacy is also known as the "fallacy of labor scarcity", "lump of jobs fallacy", a "fixed pie fallacy", or a "zero-sum fallacy". [76]
10. We could still be alive: disconnection is not death!
11. Yes, there are exceptions like Veblen goods, but this rule is true the vast majority of the time.

12. I made these using a notebook from the Wolfram Demonstrations Project: https://demonstrations.wolfram. com/DoubleSpringPendulum/.

13. Stephen Wolfram has offered a prize to anyone that can figure out some simple things about this particular Cellular Automaton: https://writings.stephenwolfram.com/2019/10/ announcing-the-rule-30-prizes/.

14. See section *2.13, Unitarity – Unitary Evolution* for an explanation of why information is never lost.

15. See section *2.13, Unitarity – Unitary Evolution* for an explanation of information conservation within quantum mechanics.

16. Presumably the choice is made by a conscious universe with intelligence in its every pore! In truth, the Copenhagen interpretation is such utter unmitigated codswallop that it is easy to keep making fun of it. Niels Bohr should have been totally embarrassed to have proposed it and his acolytes should have been (and should be) ashamed of themselves, but that is another story.

Chapter 4

Free Will in Practice

Having made the distinction between FWIT and FWIP, what does it mean to make a choice in FWIP? In one sense, it means *exactly what it means without making the distinction between FWIP and FWIT*. You decide to go for a walk. You decide to eat your steak medium rare. You decide to buy an oven. It's all the same. Nothing practical has changed. One can go about life exactly as before.

But in another sense, many things have changed. Once you realize that everything is following the inexorable and deterministic laws of physics, you realize that, despite your inability to predict the future, whatever was going to happen is what happened. Your reaction to reading this book—disbelief, annoyance, or the thought that the author has lost his mind—all of those are running on a deterministic substrate which means that what was going to happen is what is happening *including what is going on in your brain!* And, yes, it means that from the Big Bang onwards, everything is happening according to a deterministic set of rules, which means that the *ultimate cause*, but not the *proximate cause*, of absolutely everything was set at the moment of the Big Bang.

This is a very important distinction which is often lost in discussion. Practically, it is perfectly correct to say that Person X committed crime Y, or that driver A's carelessness hurt driver B. Those are all *proximate causes*. And, as proximate causes, they are what humans have been, are, and always will be, concerned with. It is perfectly correct to argue that economic policy A is better than B or that political structure P is worse than Q. Again, those are all proximate causes. The fact is, however, that those are not *ultimate causes*. It's a question of which level of explanation you want. That the ultimate cause is the Big Bang isn't going to help you get food on the table, pass an exam, or write a scathing letter to your senator decrying their lack of intelligence.

Making a choice in FWIP means that, at a given point, more than one outcome is possible *and* that only one of those outcomes is going to happen.

4.1 Many Worlds

Stealing a leaf from David Deutsch, let's think about why the earth has seasons. Consider someone living in Greece in antiquity. This person came up with the axis tilt theory of the seasons. The axis of the earth is tilted and when a specific hemisphere of the earth is pointing towards the sun, it is summer there. When it is pointing away, it is winter. This theory will have predictions. Despite having never been more than a few miles from where you were born, you can confidently say that as you keep going north from Greece, the weather will get colder and colder in the winter until you reach a point where in the winter the sun never rises and in summer the sun never sets. As you go south, you reach a point where there are basically no seasons (the equator) and as you keep going south you find that the seasons are now reversed. Summer in the southern hemisphere is winter in the northern and vice versa. It's a good theory because it is *both testable and hard to modify*. Now people make fun of you for making such wild predictions so you consider modifying this theory. But this is very very very hard.

Yet because it is a good explanation—hard to vary—it is not yours to modify.... You cannot confine its predictions to a region of your choosing. *Whether you like it or not, it makes predictions about places both known to you and unknown to you, predictions that you have thought of and ones that you have not thought of.* [emphasis added] [43].

This is the situation we have with the Many Worlds interpretation of quantum mechanics. The Schrödinger equation is unambiguously deterministic. There is no "collapse" of the wave function. That is something added to the theory from the outside, to make it conform to what we think is "sensible". The theory itself is perfectly clear: when an interaction takes place, no possibilities are lost—all possibilities do occur. What we experience is simply one branch of all those possibilities, the one that leads to us today, eating chocolate bonbons while lying on a couch reading this book. But there are many other branches where other choices are made that did not lead to us lying on a couch eating chocolate bonbons and wondering why on earth we're reading this book.[1]

> Thus the reach of an explanation is neither an additional assumption nor a detachable one. It is determined by the content of the explanation itself. The better an explanation is, the more rigidly its reach is determined—because the harder it is to vary an explanation, the harder it is in particular to construct a variant with a different reach, whether larger or smaller, that is still an explanation. [43].

The simplest and cleanest way of thinking about quantum mechanics is via the Many Worlds interpretation. Every other interpretation requires the equivalent of epicycles.[2] The burden of proof (i.e., that the other interpretations are correct) is on those proffering those alternative explanations, not on the Many Worlds interpretation because it isn't really even an "interpretation"—it is *literally* what the quantum mechanical equations are saying! And there's nothing particularly mysterious here either. The Many Worlds interpretation isn't saying that there are universes constantly splitting off. That's the cute, sensationalist and titillating (to the extent that physics can titillate!) statement. In reality, every time a choice is made, the wave function *of the universe* splits into as many branches as there are choices, with the "thickness" of each branch being the probability of that branch occurring *and* those branches remain disjoint forever—they never interact again. Human language finds it hard to explain that, so we call each branch a "world" but there really is only one world: our universe. Our universe has only one wave function: the wave function of the universe. All of our human activity, all the motions of the stars and planets, supernovas, black holes, and every other thing the universe "does" is part of that *one* wave function. Our attempts to understand the universe cause us to deliberately modify that wave function so that we can understand something about its constituent parts. One way to think about this in ordinary language is that the universe "runs lots of computations in parallel". Our experience is that of living on one branch of those parallel computations. Each time a choice is made there is a new "branch"—more parallel computations take place. Once those branches split, they do not ever rejoin. Maybe "Many Worlds" should be called "Many Histories". Also, for the technically minded, I want to dispose of one objection here. Many physicists believe that Bell's theorem rules out any theory with "hidden variables". This parallel computation way of thinking can be construed as a "hidden" variable theory. What confuses nonspecialists in the foundations of quantum mechanics is that Bell's theorem only rules out "local" theories with hidden variables. It *does not* rule out non-local theories with hidden variables. "Non-local" in this context means that parts of the universe that are outside the light cone of the events we are observing can affect the observation we are making. See section *2.20, Bell's Theorem.*

4.2 Newcomb's Paradox

Imagine two boxes, A and B. A contains $1000 and is clear so you can see inside. B is opaque and you can't see inside.

There is another person called the "predictor", who is a reliable predictor. There is also you, who is the "player". You will be allowed only one of two choices: either take just box B *or* take both box B & box A. The predictor, being a reliable predictor, predicts (in their own mind) which box or boxes you, the player, will choose, *in the future*. Having made that prediction, the predictor loads box B with nothing, if they predict that you will take both boxes, and with $1,000,000 if they predict you will take only box B. Now the question is: *which choice should you make as a player?* There are four possible outcomes.

- Predicted A+B, Actual A+B for a payout of $1000.
- Predicted A+B, Actual B for a payout of $0.
- Predicted B, Actual A+B for a payout of $1,001,000.
- Predicted B, Actual B, for a payout of $1,000,000.

Approximately half of the people to whom this thought experiment is shown choose taking both boxes. That is because there is no condition under which you do not get a payout. If the predictor was wrong, you get $1,001,000 and if the predictor was right, you get $1,000. On the other hand, the other half say that if the predictor is reliable, then you should choose box B since the predictor would very likely know you had chosen box B and so you'd get $1,000,000 with a high probability.

Now imagine if the predictor was not just reliable but perfect. In this case you must always take box B. The two possibilities where the payouts are $0 and $1,001,000 both require the predictor to have made a mistake. Since, by construction, the predictor is perfect, those two are ruled out. From the two that are left then, the better choice is taking just box B for a payout of $1,000,000. Here then is the paradox for our purposes (internal citations omitted):

William Lane Craig has suggested that, in a world with perfect predictors (or time machines, because a time machine could be used as a mechanism for making a prediction), retrocausality can occur. If a person truly knows the future, and that knowledge affects their actions, then events in the future will be causing effects in the past. The chooser's choice will have already caused the predictor's action. Some have concluded that if time machines or perfect predictors can exist, then there can be no free will and choosers will do whatever they're fated to do. Taken together, the paradox is a restatement of the old contention that free will and determinism are incompatible, since determinism enables the existence of perfect predictors. Put another way, this paradox can be equivalent to the grandfather paradox; the paradox presupposes a perfect predictor, implying the "chooser" is not free to choose, yet simultaneously presumes a choice can be debated and decided. This suggests to some that the paradox is an artifact of these contradictory assumptions. [150]

However, as is immediately clear by inspection, since we have perfect FWIP (even while we have no FWIT) we cannot have Newcomb's paradox. While we may be running on a deterministic substrate, we cannot in fact predict anything with perfect accuracy, *and we will not ever be able to predict anything with perfect accuracy*. And so, due to the three limits of prediction (see section *4.5, Prediction Limits*) Newcomb's paradox does not occur — there is no way of ever creating a perfect predictor in our universe. And therefore, the paradox is not real, at least in our universe, which is more than good enough for me.

4.3 Consciousness
One reason that thinking about consciousness messes everybody up is that it is, in fact, impossible for two people

to communicate their precise thoughts and experiences to each other. When we attempt such a thing we use an approximation: language! But language is at its very core a representation of only a few salient features of the speaker's experience—that portion of experience that can be encapsulated in words. This is why, for example, it is hard work being an author. It's not as if an author can simply stand in front of a machine and allow his or her thoughts to come spewing out at the speed at which he or she thinks them. Even gonzo journalists like Hunter S. Thompson had to work to make it seem so! Those thoughts have to be organized and expressed in a manner that they can be understood by someone else. And yet, as we all know, reading someone else's thoughts is at best a very pale imitation of those thoughts.

The literature on consciousness has this quality to it. It is very much like the blind men feeling an elephant for the first time. One feels its trunk, the other feels its leg, the third feels its tusks. What is happening is that each person describing consciousness is describing some particular aspect of their experience of what they think consciousness is. It is no wonder then that everyone's descriptions of this particular experience that we all have in common are so vastly different.

Or think about words. All of us use the same words when speaking the same language. Yet, quite frequently, we mean different things when we use the same words in the same way. When we write we are never quite sure what the reader is going to interpret from what we have written. One interesting example of this comes from law. In patent law there is a process called a Markman hearing. [149] In a Markman hearing, which happens during patent infringement lawsuits, the judge has a hearing to determine what the words in the patent mean. Now, as a practical matter, this is an absolute necessity. Yet if you think about this from a theoretical basis, it is an absolute absurdity! The words mean whatever the words are supposed to mean! We should all be able to agree on what they mean! Yet we don't!

This problem is overcome in computer languages. In any computer language we are using standard English words but we are using them in very precise and restricted ways. In normal language or speech it is frequently unclear what any given word is supposed to mean. A digital computer cannot interpret that kind of ambiguity and needs exacting input (else "garbage in, garbage out"). For precision then, we create a highly restricted language with exact meanings using English words in highly specific ways. That specific collection becomes the computer language. And there are hundreds of popular computer programming languages—all of which are equally precise, but in their precision, they leave behind flexibility. And because they leave behind flexibility (i.e. words must mean precisely what they mean and nothing more or less) we have to have different computer languages that are tuned for different tasks—there isn't a single computer language that can do everything—each has its strengths and weaknesses. And, as a corollary, this is also why the whole world now programs in English. Yes, it is true that the beginning of computer programming was very English focused and that helped the spread of English in programming. However, there was nothing stopping other countries from adopting their own languages to program computers. A computer language is compiled to machine code anyway: 0s and 1s. So it *really* does not matter what language you program in and yet everyone uses English. Universally, everyone adopts the highly restricted English that leaves no ambiguity when programming a computer. This even extends to spelling within countries that share the same language. In British English "program" is spelled "programme". But not in British computer programming!

Also, let us not confuse free will with consciousness. We have a very hard time creating a definition of consciousness upon which everyone can agree. I'm not arguing that unpredictable things are conscious. I am arguing that unpredictability is the defining feature of free will. Whatever consciousness is, it's not the same thing as free will. Exhibiting free will does not mean the exhibitor is conscious. In that sense humans are no different than an unpredictable algorithm. Because given the determinism of the laws of physics that's exactly what we are: a very very very[3] complicated arrangement (algorithm) of atoms all obeying the laws of physics.

A simple thought experiment serves to convince that there is nothing terribly "mystical" about consciousness (or intelligence for that matter). The brain's "thinking" apparatus consists of neurons. A neuron has some incoming connections called "dendrites" and (usually) a single outgoing connection called an "axon". Neurons connect to each other with axons connecting to dendrites across a tiny space called a synapse. Neurons send signals down their axon by sending a chemical (a neurotransmitter) to the dendrite of another neuron. A neuron fires when enough of its dendrites receive an incoming signal— when a certain threshold is reached, the neuron fires and sends a signal down its axon. The neuron can either fire, or it can not fire. Those are its only choices. With that in hand, we can consider the following. Suppose that there is a person called Jack and that we all agree that Jack is a human. Suppose now that we take one of Jack's neurons and create a precise model of it such that we know exactly what inputs produce what outputs. Next, we create an artificial neuron that matches that model and then replace the original neuron with this new artificial one. This is perfectly possible—there is nothing in the laws of physics that prohibits this—it's just difficult to do with our current technology. After the replacement, is it still Jack? Under any reasonable definition of "Jack", yes it is and must be. And now you know the drill. Replace a second neuron, then a third and so on. Is it still Jack? Eventually, you will have replaced all the neurons in Jack's brain with new, artificial, neurons. Is it still Jack? Unless you want to posit that there is a "phase transition" of some sort that occurs when enough original neurons are replaced by exactly equivalent artificial ones, you are left with the (possibly uncomfortable) conclusion that artificial Jack is just as much "Jack" as original Jack was. And thus, we have now done several things. One, we have constructed an artificial brain that is still "Jack"! Two, if "original" Jack before replacement was conscious, so is "artificial" Jack after replacement. And three, we have conclusively demonstrated that there is nothing terribly special about Jack's brain *except for the precise arrangement of its neurons—that is to say, Jack's "connectome", which is the full list of all the connections between all of Jack's neurons!* This thought experiment also shows, conclusively, that an "artificial" intelligence could be both intelligent and conscious because we have just constructed an entirely artificial being that is conscious and human!

Determinism means that there is nothing mysterious about consciousness. It is simply the result of a very large number of deterministic processes interacting with each other in an unimaginably large (but finite) number of ways. And, as new research is beginning to show, it is likely an even more complicated collection of deterministic processes interacting with each other than we originally thought because it is becoming increasingly clear that the brain and body are inextricably linked and that the body influences the mind and vice versa. [36] Consciousness, whatever it may be, is an emergent property; it has to be since there is precisely no other possibility. Given that we are an arrangement of standard atoms, consciousness is a necessary outcome of the deterministic physics of our universe!

4.4 Universality

The decimal digits of irrational numbers like π cannot be predicted out of sequence via any algorithm. They can only be laboriously calculated by means of specific algorithms.

No shortcut is possible as those digits provably exhibit no pattern. On the other hand, rational numbers such as 3/2 = 1.5 or 4/3 = 1.333333... either have a decimal representation that stops or repeats forever (the ... after the 3 means that the 3 repeats forever). But as we have seen, free will is the result of unpredictability. So does π have free will while 3/2 and 4/3 do not? In one sense, yes. There is no predictability whatsoever in π. It doesn't personally bother me at all to call that unpredictability "free will in practice" the same way as it doesn't bother me at all to think of the weather as having free will in practice (i.e., a mind of its own). However, I suspect that this is an unsatisfying answer because it would imply that humans and π are the basically same thing. I suspect, therefore, that a truly satisfying explanation of free will in practice also requires universality.

As we have seen, even simple systems can display universality. And that means that even simple systems can be computationally irreducible and thus are unpredictable. If something is predictable then it is not universal. So while a sense and respond creature like a bacterium is indeed running on a universal substrate (the laws of physics), its precise arrangement of those physical things (atoms) means that it is unlikely to be universal. So it likely does not have much, if any, free will. A dog is less predictable than a bacterium. That part of the dog that is predictable we might term instinct. The term "instinct" is just another way to refer to that portion of behavior that is predictable. A dog then is partly instinct and partly computed behavior: it has more free will than a bacterium. A dog is also substantially less predictable than a bacterium. That lack of predictability suggests that the arrangement of atoms that makes up a dog is an arrangement that allows universality. The human brain is of course an incredibly complex system, far more complex than simple systems, and so is universal. And since the brain is universal, it will be computationally irreducible, and since it is computationally irreducible we have will always have free will in practice.

My guess is that the division amongst creatures that have "free will in practice", and those that do not occurs at the point at which you get creatures that are not acting solely on behavior whose programming is predictable. The moment you get a complex enough creature such that its arrangement of atoms allows for universality, you get free will in practice. This neatly differentiates dogs and humans from π because the decimal digits of π are not universal. I propose that we use the presence of universality to divide those things that possess a kind of free will that we colloquially agree really is free will in practice, from those that we colloquially agree do not possess free will in practice even if otherwise meeting the criterion of unpredictability. This isn't necessary for the truth or falsity of my overall argument, but it is a nice and neat division of two groups.

4.5 Prediction Limits

Thus, we come to the most important point about FWIP. FWIP is synonymous with unpredictability. *There are three different limits to prediction: the practical limit, the chaos limit, and the irreducibility limit. Each one of those is individually impossible to overcome, but in fact, we have three!* One could certainly argue that the first two, the practical limit and the chaos limit, are both, in some very broad and general sense, technological. And, if one is an über optimist, one could argue that as long as there is no law of physics against it, eventually, it must happen. And so perhaps the first two limits are not necessarily dispositive. I don't believe that is true, and I do not think either can ever be overcome. But you don't have to take my belief for it because the third limit — the computation limit — is fundamental and *cannot* be overcome. *We have massive redundancy in the limits to prediction: limited predictability is highly fault tolerant!*

4.6 Fruit Flies

As an example, let's consider fruit flies. Fruit flies live about two months and are tiny: about the size of a poppy seed. But even this tiny creature has a very complicated brain. Its brain contains 3016 neurons, with 548,000 connections between them. It took scientists at the University of Cambridge 12 years to map all these connections. In a similar vein, it took scientists 50 years to map the brain of a roundworm, a creature that has only 302 neurons. So now we know, with absolute certainty, the exact connections and layouts of fruit flies and roundworms. [115] We can, if we wish, simulate these in silico. So do these creatures have free will? And this is where all the issues from prediction limits come into play. First, to know what a fruit fly or a roundworm is going to do before it does it, we have to simulate it faster than its own brain can compute *and* we must simulate its environment as well—at least that part that impinges upon the creature during the period of simulation. This is already a tall order, unless we set up an incredibly precise experiment in which the creature's environment is simplified to the fullest—and even then, we will not be able to go down to the molecular level of all the air molecules that impinge upon it during the experiment. Second, with that many connections, there will a great number of nonlinearities in the simulation. So both the practical limit and the chaos limit are going to show up. But even if we ignore that, unless our simulation is incredibly fast and is taking place in completely isolated conditions so that the environment has little effect on the creature we are simulating, we will run into computational irreducibility. So even in this situation, where we know everything there is to know about the system that controls the creature, we will still be unable to figure out what it is going to do until it has done it!

Notes

1. A book that was probably written, we muse, by a raving lunatic in the middle of a gigantic hallucination. Too many bad acid trips, we ask?

2. In the "Western world", before Copernicus's heliocentrism and Kepler's laws of motion, the sun was held to circle the earth in a perfect circle. Unfortunately, this produced incorrect predictions. So, to fix those predictions, additional circles were added, superimposed on each other: the "epicycles". As long as you added together a sufficient number of epicycles, you could get accurate predictions of the sun's motion. Later on, once Copernicus had argued that the earth revolved around the sun, and Kepler showed that the motion of the earth around the sun was an ellipse, epicycles fell out of favor. The reason epicycles worked despite being entirely incorrect is because, as Joseph Fourier showed, any smooth curve can be approximated to arbitrary accuracy by a sufficient number of epicycles.

3. Did I say very? Do you have a feeling of déjà vu?

Chapter 5

Consequences

Predictions are interesting things in and of themselves. For what is a prediction? It is no more and no less than a specific pattern of firing neurons in, say, my brain, which when communicated to you via these words leads to another set of neuronal firings in your brain. If you understand what I mean, then the meaning that your pattern encodes will be similar to mine. If you do not understand what I mean, or only partially understand it, then your pattern will be less and less similar to mine. If we could communicate with each other perfectly, then there would exist an isomorphism between your patterns and mine. As our communication gets less accurate (i.e., more noisy) the isomorphism disappears and we are left with an approximation. You understand some of my meaning and vice versa. The human brain turns out to *be* a prediction machine: it is always predicting what is about to happen and then it compares its predictions to its subsequent perceptions to make adjustments. *Your perception very much depends upon what you were expecting.* The Invisible Gorilla experiment is an excellent example of this. [27] The human brain evolved to ensure survival and procreation. That evolution means that it is incredibly quick at inferring patterns even when none exist. It is safer to run away at the first sound of a roar rather than to stick around to figure out if that roar was from a lion. The brain did not evolve to think about philosophy, physics, mathematics, logic, engineering, reading, writing, and so on. These are things that its pattern recognition engine can, after great effort (also called education), be taught to recognize. It is also the reason why counterintuitive truths are so difficult to accept (because they don't fit our evolutionarily evolved patterns), and also why the type of thesis I am proposing is so hard to believe. With the caveat that it is difficult to make predictions, especially about the future, let's consider some possibilities.

5.1 Prediction Zero: The Future of Determinism

No process, of any kind, living or nonliving, organic or inorganic, galactic, or microscopic, will ever be found that is nondeterministic at its underlying core. Physics is not finished and, except for fundamental physics, cannot and will never be finished. Because of computational irreducibility there will always be an infinite number of things to figure out. But, each of those will always be found to obey physical laws, which are deterministic at their core—i.e., every discovery or new physical law will reduce to a very complicated combination of fundamental physical laws. If you don't like prediction zero, call it an "assumption" but I think it is on incredibly solid ground since it is congruent with all our current knowledge—which, I remind you again, means we *know the physics that underlies all of chemistry and biology!*

Once we know (assume if you prefer) that all physical theories, both current and future, will be deterministic, we can make some more predictions.

5.2 Prediction One: The Future of Free Will in Practice

As long as the substrate is deterministic and quantum mechanics holds, no physical theory will ever exist that contradicts FWIP. This is because, as explained above, the computing power required to simulate or calculate human behavior from basic physics is forever going to be too large to be contained within the universe. Additionally, even if that were to be somehow overcome, both irreducibility and chaos will rear their ugly(?) heads. Each of these three independently guarantees that FWIP will always exist.

5.3 Prediction Two: The Shrinking of Free Will in Practice

However, the domain of FWIP will shrink, at least somewhat. Because of computational reducibility, and because of our ability to make *approximate* predictions that are good enough for most practical purposes, in some circumstances we will be able to, at least approximately, predict human behavior before it occurs. For example, FMRI experiments demonstrate that we have made a decision before we are aware of having done so: "[T]he earliest time at which decoding was possible was ~7.5 seconds (time-bin 4) before the decision was reported to be consciously made". [19]

5.4 Prediction Three: The Future of AI

At some point, as Artificial Intelligence becomes more and more widespread, we are going to be entirely unable to understand what an AI is doing, and why it is doing it. The advent of ChatGPT, chat.openai.com, which had the fastest scale up to 100 million users in tech adoption history, has made this incredibly clear. No one, not even the developers, "understands" what ChatGPT is doing. It is in the tradition of "statistical" AI where the learning is implicit. Contrast this with "explicit" AI, where the developers have programmed in rules, rather than discovered patterns in data. One example of that is the "computational engine", Wolfram Alpha, https://www.wolframalpha.com, which is the assistant that helps Apple's Siri answer questions. Wolfram Alpha is explicitly programmed via rules.

To understand what a "statistical" AI is doing would require one of the following. The first possibility is that it explains itself to us. The problem is, the way an AI works and the sheer quantity of data that it uses make it impossible for the AI to properly explain its reasoning. Humans are barely able to explain to each other why they did what they did. For an AI, it would be exponentially harder. The second possibility is that we study the AI ourselves to find pockets of "computational reducibility". The problem is that finding that reducibility is itself irreducibly hard. And the last possibility is that the AI finds in itself, or uses another AI to find in itself, pockets of computational reducibility and explains those to us. That is probably the best way forward. But that runs into the issue that the vast majority of what the AI is doing won't be reducible. In that case, how will the AI produce a useful explanation? The short answer is—it cannot. The very fact that ChatGPT "hallucinates" tells you that this approach will be fraught with difficulty. And, as an aside, consider that if the AI has large pockets of computational reducibility, we might as well not have the AI: we can then replace it with the "reduced" methods that we or it have found. So the key point is this. Only to the extent that an AI doesn't have large pockets of reducibility, will it be uniquely useful to us. But, simultaneously, the more useful it is, the harder it will be to understand what it is doing and why. *This is a necessary consequence of the deterministic paradigm.*

David Deutsch makes an argument to the contrary and argues that:

> Most advocates of the Singularity believe that, soon after the AI breakthrough, superhuman minds will be constructed and that then, as Vinge put it, "the human era will be over." But my discussion of the universality of human minds rules out that possibility. Since humans are already universal explainers and constructors, they can already transcend their parochial origins, so there can be no such thing as a superhuman mind as such. [...] Nor will artificial scientists, mathematicians and philosophers ever wield concepts or arguments that humans are inherently incapable of understanding. Universality implies that,

in every important sense, humans and AIs will never be other than equal. [43]

This assumes that *by interfacing the AI* to the brain, we will be able to speed up our thought processes:

There can only be further automation, allowing the existing kind of human thinking to be carried out faster, and with more working memory, and delegating "perspiration" phases to (non-AI) automata.

But this optimistic argument does not, unfortunately, hold. The reason is that *tasks offloaded by us to the AI will not be able to be understood by us!* Suppose that we train a neural net to simulate the movement of a human arm. We then interface that arm with someone who has lost their limb. Let's assume that the interface is perfect and the arm does what it is asked to do by the brain of the human to which it is attached. Can that human ask the arm: "How did you do that?" The answer is obvious: no. In fact, consider the following: move your leg, or wiggle your toe, or type a letter, or hit a golf ball. Now, please ask yourself to explain how you did it. The answer is that, in general, you cannot. It takes great effort to explain "how" you do anything: this is one reason sports coaches are paid as much as they are—there is a skill and an art to being able to describe what to do in a way that is understood! And even then, the person being coached has the difficult task of interpreting the coach's instructions in order to carry out whatever task they're trying to do. Additionally, there is a lot of evidence to suggest that the better we get at doing something, the worse we get at being able to explain it. That's because the knowledge is "implicit": it is encoded in a pattern of neural firings in our brain. It isn't encoded in a way that is easy to turn into language to communicate to someone else.

Even further, quite often we do not even know why we did something! As David Brooks puts it:

> One of the most unsettling findings of modern psychology is that we often don't know why we do what we do. You can ask somebody: Why'd you choose that house? Or why'd you marry that person? Or why'd you go to graduate school? People will concoct some plausible story, but often they really have no idea why they chose what they did.
>
> We have a conscious self, of course, the voice in our head, but this conscious self has little access to the parts of the brain that are the actual sources of judgment, problem-solving and emotion. We know what we're feeling, just not how and why we got there.
>
> But we also don't want to admit how little we know about ourselves, so we make up some story, or confabulation. As Will Storr writes in his excellent book *The Science of Storytelling*, "We don't know why we do what we do, or feel what we feel. We confabulate when theorizing as to why we're depressed, we confabulate when justifying our moral convictions and we confabulate when explaining why a piece of music moves us." [20]

This issue is going to crop up as AIs become more and more complex and capable. As the rate of computation speeds up and the capabilities of computers improve, tasks requiring decisions are going to be outsourced to AIs. Take a self-driving car for instance. Can you ask it a question and expect a clear answer? Even with what are (compared to hypothetical AI programs of the future) simple AI programs, it takes enormous effort, creativity, and deep knowledge to be able to debug an AI.[1] Interfacing with the AI isn't going to solve that problem. All the interface does is increase the rate at which information can be transferred to and from the human via the interface. *It assuredly does not mean that we will understand what the AI is doing!* Just think of the fact that ChatGPT is able to do things that it was never programmed or intended to do—in other words, it has latent capabilities. For example, it can play chess, and play it quite decently at least for the opening moves! [83] And, ChatGPT is showing that it is developing its own "theory of mind": the cognitive ability to understand and attribute mental states—like beliefs, intents, desires, emotions, and knowledge—to oneself and others, and to understand that others have beliefs and perspectives different from one's own. [134]

We already have opaque programs being used by law enforcement for the purpose of "risk assessment" with regard to the probability of recidivism. [66] As Stephen Bush put it:

> The most worrying use of algorithms in policy are so-called "black box algorithms": those in which the inputs and processes are hidden from public view. This may be because they are considered to be proprietary information: for example, the factors underpinning the Compas system, used in the US to measure the likelihood of reoffending, are not publicly available because they are treated as company property. [...] The other form of black box algorithm is one in which the information is publicly available but too complex to be readily understood. This, again, can have dire implications. If the algorithm that decides who is made redundant cannot be reasonably understood by employees or, indeed, employers, then it is a poor tool for managers and one that causes unhappiness. In public policy, if an algorithm's out-workings are too complex, they can confuse debate rather than helping policymakers come to better decisions. [23]

Another example is from the experience of Sarah Bell, an AI specialist that focuses on real estate. She found that the AI she was testing was producing results that were biased against Australians! And, even more worryingly, it was not possible to figure out why the system disliked Australians because it was effectively a black box. [133] These programs are already causing angst, and yet *they are extremely simple compared to what is coming in the future!*

Think of communicating with a dog. Some breeds of dogs have over 100 million scent receptors in their noses. Humans, by contrast, have only 5 million. One-eighth of a dog's brain is devoted to smell and dogs can detect smells at concentrations as low as a few parts per trillion. And it isn't even that their sense of smell is more acute than ours. It is also that they can smell things that we simply cannot—they smell complexity where we smell simplicity.

Dogs can detect some smells in parts per trillion. However, extra scent receptors don't just mean dogs can sniff subtle odors we would miss. They also allow dogs to detect a complexity in odors that humans can't. You might smell chocolate chip cookies, but your dog can smell the chocolate chips, flour, eggs, and other ingredients. And when dogs sniff another dog, they smell more than doggy odor. They can detect the gender of the other dog, as well as clues to that dog's age and health status. No wonder dogs find "pee-mail" on the fire hydrant so fascinating. They're getting all the neighborhood gossip in one big whiff.

Given that, suppose we (somehow) learn to communicate with a dog. How is a dog supposed to describe a scent to us that we are simply unable to experience or comprehend? For example, dogs can *smell* time! [46] And yet, despite that, they are "present focused" — they have no sense of tomorrow. So how do you tell a dog that you will take him for a walk tomorrow? Or tell him about something that happened a week ago or a month ago? Furthermore, dogs communicate via pheromones as do humans. But it is extremely likely that their signaling is very different than ours given the entirely different makeup and capabilities of their scent and vomeronasal organs.[2] Since we don't have the same scent capabilities and we most certainly do not have their vomeronasal organ, how will they communicate their thoughts about what they are smelling? And so, *if we can't communicate with "man's best friend", how on earth do we think we will be able to successfully do it with an Artificial Intelligence, running at speeds we cannot comprehend, ingesting quantities of data we cannot imagine, and with an algorithmic complexity that exceeds anything we have experienced?*

Just think of the difficulties when describing wines! For example, *The Underground WineLetter* has some award winners for what it calls "stupid wine descriptions". Here are some excerpts:

Initially, the nose is very closed, but it eventually opens to reveal lovely aromas of beeswax, wet wool, dried pear and honeycomb. The palate is similarly dumb at first, but it unfurls with a delicate, citrus entry, a carefully planned crescendo of dried pineapple, lanolin and spice, with hints of hazelnut on the finish.

A 1 hour decant and a couple of muscular swirls revealed a rich garnet color and an exquisite nose of blackberry and rich vanilla Devon custard. Leonetti's use of Merlot and Cabernet Franc rounds out the 80% Cabernet Sauvignon and adds that subtle minty dill pickle tang on the nose so characteristic of the producer [...]

Full-bodied, concentrated and downright explosive in the mouth, it is still wearing loads of gorgeous puppy-fat fruit flavors [...]

Quite dry, juicy and elegant, featuring stone fruit and mineral flavors sexed up by a flinty nuance on the end. [135]

And this is *humans* describing taste and scent to humans!

We can attempt to program AIs to always explain themselves as best they can. I suspect this will prove to be unsatisfactory at best, and completely impossible at worst. And we can try to make sure that AIs are always constrained in some way or ways. This is likely somewhat more promising. Isaac Asimov's three laws of robotics [155] were an early attempt to wrestle with the problem of what to do with an artificial intelligence that humans are no longer equipped to control or constrain. They are worth quoting in full:

1. A robot may not injure a human being or, through inaction, allow a human being to come to harm.
2. A robot must obey the orders given it by human beings except where such orders would conflict with the First Law.
3. A robot must protect its own existence as long as such protection does not conflict with the first or second law.
4. Asimov eventually added a zeroth law: A robot may not harm humanity, or, by inaction, allow humanity to come to harm.

These laws are an outstanding attempt to constrain AI in such a way that it "works" for humans. But a moment's reflection will immediately reveal that even following such programming, a robot could do things of which no human would approve. A version of the trolley problem [156] is an excellent example. If a robot has to kill one person to save fifty, what should it do?

Then there is the effectiveness issue. The more we constrain an AI, the less effective it will be. The issue is that we need to allow the AI to explore its parameter space for maximum effectiveness: the more combinations it explores, the more effective it can become. Therefore, except for specific cases where the constraints somehow make it more efficient (perhaps by stopping it exploring those configurations that we know we do not want) this *must* be true: the *more* we constrain it, the *less* effective it will be. So where will we draw the line?

Roger MacBride Allen explores what happens when you constrain robots in his book *Isaac Asimov's Utopia*. [10] To avoid a spoiler, suffice it to say that the combination of robots following the three laws above, robots that follow a modified version of the three laws, a robot that follows no laws, and humans produces incredibly unexpected, but highly logical, results. And the conclusion of the book is completely in line with the argument of this section.

Next there is the issue of explanation. As AIs become more capable, asking them to explain themselves may become impossible. In many ways, this is already happening. One of the most pressing issues in molecular biology, for the last 50 years, has been protein folding. Even when we know the chemical structure of a protein—what atoms it is made of—we usually do not know what shape that protein has. This is incredibly important because proteins support essentially all functions of life. But what a protein does depends mostly on its 3D shape— how it is "folded"—and not so much on what atoms it is made of. For 50 years, some of the smartest biologists have tried to crack the problem of figuring out how a given protein would fold and have failed.

We have still not unravelled all the rules that govern how a protein folds reproducibly into a particular shape, specified only indirectly by its DNA sequence. *Ironically, artificial intelligence algorithms have recently made some progress on this question, but we're not quite sure how they did it.* [emphasis added] [86, p. 13]

And yet! In 2020, DeepMind, using the latest large AI models, "released AlphaFold protein structure predictions for nearly all catalogued proteins known to science." [42] So now, ask yourself, how will DeepMind explain to us what it did, and how it did it? It's impossible—it has way too much data as input, its internals are incredibly complicated, and there is no obvious "reducibility" present for it to exploit such that we can understand what it did and how it did it. *There is no conceivable way in which DeepMind could communicate with us in a manner that would be useful to us, and that we could understand.*

The lack of explanation from an AI is going to become a serious problem. In particular, an AI may produce results that we consider biased. But, given that the AI has access to exponentially more information than any human, it will be impossible for us to know if the bias is justified or is what we would consider ridiculous. As it is, humans display truly bizarre biases and yet, because we can ask them for explanations, we feel reasonably comfortable with letting them make decisions. For example:

Employing the universe of juvenile court decisions in a U.S. state between 1996 and 2012, we analyze the effects of emotional shocks associated with unexpected outcomes of football games played by a prominent college team in the state. We find that unexpected losses increase sentence lengths assigned by judges during the week following the game. Unexpected wins, or losses that were expected to be close contests ex ante have no impact. The effects of these emotional shocks are asymmetrically borne by black defendants. The impact of upset losses on sentence lengths is larger for defendants if their cases are handled by judges who received their bachelor's degrees from the university with which the football team is affiliated. Different falsification tests and a number of auxiliary analyses demonstrate the robustness of the findings. These results provide evidence for the impact of emotions in one domain on decisions in a completely unrelated domain among a uniformly highly educated group of individuals (judges) who make decisions after deliberation that involve high stakes (sentence lengths). They also point to the existence of a subtle and previously-unnoticed capricious application of sentencing. [53]

If AIs cannot provide us explanations, then what about consent? For example, India is relocating 100 cheetahs from South Africa to India as "part of an ambitious effort to reintroduce the spotted cats in the south Asian country." [73] Let's put aside any issues with unintended consequences. Did anyone ask the cheetahs for their consent to be uprooted from their home and moved to a completely different continent? That sounds absurd, and it is. The issue is that we cannot communicate with the cheetahs, and even if we could, it is entirely unclear that we would be able to explain to them, in *language that they will understand*, what we want to do, why we want to do it, and why it benefits them. Is it that we feel that the vast gulf of intelligence between us and the cheetah allows us to make decisions for its own benefit, regardless of what it thinks? If so, what's to stop an AI from making such decisions for us? And what would we do about it?

Consider the following argument, put forward by Katja Grace. [59] Why don't we trade with ants? There are all kinds of useful things ants could do for us: cleaning things we find hard to reach, building various kinds of structures, killing insects that are harmful to humans, digging tunnels, fixing things particularly in tight spaces, and so on. We could easily pay them in food or any other commodity that they value. This is not just idle speculation on my part. For example, recent experiments have shown that ants are able to sniff out cancer in rodent urine and can be more quickly trained than dogs. [12] So why don't we trade with them? The answer is, we cannot communicate with them! We can't explain to them what we want, and they can't consent to trade with us to provide what we want in exchange for what we can provide to them.

There are interesting parallels between how we deal with AI and how we deal with human experts. Unless you're a physicist, you're unlikely to know much about quantum mechanics and quantum field theory. So how do you, dear reader, know that what I'm telling you is correct? At some point, you have to throw up your hands and say, well, a physicist told me that so it must be true. The same holds in law, medicine, accounting, etc. Any field in which there is a disparity of knowledge will have this problem. In many of those cases, you just have to follow what the expert says, and you do not always really know why. If you have cancer, do you know enough biochemistry to challenge the doctor's recommendation for which oncology drug to use? Or whether to try immunotherapy? If you have a complicated tax issue in front of the IRS, and your tax lawyer makes some very complicated argument to the IRS that you barely understand, do you know enough to challenge why the lawyer is making that argument? Now imagine an AI that is vastly more experienced and knowledgeable than any human. How can that AI even begin to explain its thinking? Suppose that you asked me for an explanation of quantum mechanics and you're not well versed in modern physics and the associated mathematics. How would I explain it to you in terms you'd understand? At best, all I could do was make analogies and the best you'd be able to do is get a very rough idea of what quantum mechanics is. Amusingly, the great physicist, Richard Feynman, would always say, tongue in cheek, if you can't explain something to your bartender, you do not really understand it. By that standard, no one understands either quantum mechanics or general relativity, which are the two foundations of all of physics.

The same issue comes up with games. Take, for example, the game "Diplomacy" which is a board game that encapsulates the strategic depth and psychological nuance of pre-World War I European geopolitics. It stands out because it lacks the randomness of dice or cards; instead, it's a pure test of a player's ability to negotiate, form and dissolve alliances, and subtly manipulate the game's balance of power. In this game, each player represents one of the seven major European powers of the era. Their objective is to claim territories and amass influence across the continent, with moves executed through written orders after a phase of open-ended discussion. Herein lies the game's essence: it's about predicting and influencing the human element, with victory often hinging on persuasive discussions, secretive pacts, and the occasional treacherous betrayal. The Machiavellian nature of Diplomacy is exemplified by Cicero, an AI developed by Meta. Named after the skilled Roman orator, Cicero has demonstrated a stunning capacity to engage in the game's complex social manipulation. It uses natural language processing to communicate with human players, successfully persuading and strategizing without their knowledge of its nonhuman identity. This AI's ability to deceive and strategize raises profound ethical questions. It was able to play Diplomacy online and outperformed most human participants by employing the very tactics that define the game's human challenge: negotiation, bluffing, and misdirection. Cicero was trained on vast amounts of data from tens of thousands of games, learning not just to strategize but to understand and predict human motivations, making it a formidable opponent.

We already have AIs that can lie, cheat, and strategize—albeit in a game designed for lying, cheating, and strategizing—to defeat the very best players and advance their own interests! [7]

Or think about "decompilers". When we write computer code, we write it in a high-level language such as C, Java, Fortran, C++, Python, etc. That code is then converted to low level instructions that a computer's processor can understand. This can be done in advance, producing an "executable binary file", or it can be done on the fly via "interpretation". This process is called "compilation". When you run an executable such as Microsoft Word on your computer, you are running a compiled version of high level computer code. Now suppose you would like to recover the original high level code that generated the low level computer code. Well, you can't! You can use a decompiler, and that decompiler will produce high level code in a language of your choice, but it will most assuredly look nothing like the original code that was compiled—compilation necessitates information loss! Similarly, if you ask an AI for an "explanation" you're asking it to convert what it knows into a language that you understand: it will be converting information from a form that it understands, into a form that you understand. By definition, there will be information loss. And for all the reasons mentioned above, the information loss from that conversion is going to be very substantially larger than that coming from a simple decompilation, which after all, originally *started* as human created information.[3] Now, there certainly have been attempts to create a type of machine learning that can "explain itself".[4] [167] The problem is that these attempts only work when the data isn't "noisy". So, for example, if you use these types of algorithms to figure out Kepler's Laws of Planetary Motion or Newton's Laws, they will likely be successful. There is plenty of regularity in the data that the algorithm can "latch" onto, and then explain that regularity to us. On the other hand, if you ask these types of algorithms to look at something complicated like, for example, financial data and tell you why a certain stock was an attractive buy, they would fail. There really is no good way to get a different "intelligence" that works on different principles than ours to explain itself to us in a manner that we will be able to understand.

There's still more. For example, I personally have some vague idea what an adverb is or what a "noun clause" is. But I have not the faintest idea what a gerund is, or what a past participle is, or frankly, what a participle is. I don't know what a future perfect tense is, or indeed, what on earth it is used for. I can somewhat distinguish present, past, and future tenses but not particularly well. I have the vaguest idea of the difference between subjects and objects in a sentence and quite often I have no idea why a certain word in a sentence is considered a "verb". The definitions of these things seem almost arbitrary to me. Therefore, I have essentially no clue about the rules of English grammar. However, as I pointed out at the beginning of this section, there is a difference between implicit learning and explicit learning. If you learn English grammar via rules, you are engaging in explicit learning, similar to Wolfram Alpha. If you learn English grammar via reading, writing, and using English, you are engaging in implicit learning. You can potentially explain the former, but it is very difficult to explain the latter, assuming it is even possible. Despite knowing very little English grammar in the form of rules that I can explain, here I am writing a book in English. Obviously, it's my native language. Not so obviously, it means that *I cannot teach anyone else the rules of English grammar!* They're in my head—I rarely make grammatical mistakes (or for that matter spelling mistakes)—but I couldn't possibly explain them to you. So, if I don't know *explicitly* what I do know *implicitly* then there is no possibility of my being able to explain to anyone else why some grammar is correct and why some grammar is incorrect. For an amusing and highly specific example, consider this rule:

The paragraph concerned the order of adjectives—if you're using more than one adjective before a noun, they are subject to a certain hierarchy. You know it's proper to say "silly old fool" and wrong to say "old silly fool", but you might never have thought about why—or if you did you probably imagined it was just some time-honoured convention you picked up by rote. But it isn't. There's a rule.

The rule is that multiple adjectives are always ranked accordingly: opinion, size, age, shape, colour, origin, material, purpose. Unlike many laws of grammar or syntax, this one is virtually inviolable, even in informal speech. You simply can't say My Greek Fat Big Wedding, or leather walking brown boots. [47]

That is the same position we are already in with machine learning and AI: the bigger models cannot explain how they do what they do. Those models are only going to get bigger and more complicated, and there is *no chance* that they will be able to explain to us, in terms we can understand, why they did what they did.

And lastly, there is the communication issue. Imagine, for example, that you are trying to communicate with your dog. Remember, this is "man's best friend." We don't know in any detail what a dog is thinking, *even one that has lived with us for its entire life.* So what makes us think we will know what an AI is thinking or why?

This doesn't even touch the more quotidian concerns: that AI (and more generally, all types of machine learning) could exacerbate inequality, threaten our autonomy as individuals, and threaten our privacy. And, of course, an AI can work extremely well even if we don't understand how it works. As David Weinberger puts it: "The complexity of ML models can make it difficult to debug them, to spot mistaken outputs, and to protect them from being subverted, by, say, carefully positioning a piece of tape on a traffic sign or altering a few pixels in an image." [139] Or as Jemima Kelly put it:

> The most common arguments against AI-driven adjudication, like this one, concern outcomes. But in a draft research paper, John Tasioulas, director of Oxford university's Institute for Ethics in AI, says more attention should be given to the process by which judgments are arrived at.
>
> Tasioulas quotes a passage from Plato's The Laws, in which Plato describes the qualitative difference between a "free doctor"—one who is trained and can explain the treatment to his patient—and a "slave doctor", who cannot explain what he is doing and instead works by trial and error. Even if both doctors bring their patients back to health, the free doctor's approach is superior, argues Plato, because he is able to keep the patient co-operative and to teach as he goes. The patient is thus not just a subject, but also an active participant.
>
> Like Plato, Tasioulas uses this example to argue why process matters in law. We might think of the slave doctor as an algorithm, doling out treatment on the basis of something akin to machine learning, while the way the free doctor treats his patients has value in and of itself. And only the process by which a human judge arrives at a final decision can provide three important intrinsic values.
>
> The first is explainability. Tasioulas argues that even if an AI-driven algorithm could be programmed to provide some kind of explanation of its decision, this could only be an ex post rationalisation rather than a real justification, given that the decision is not arrived at through the kind of thought processes a human uses.

The second is accountability. Because an algorithm has no rational autonomy, it cannot be held accountable for its judgments. "As a rational autonomous agent who can make choices ... I can be held accountable for these decisions in a way that a machine cannot," Tasioulas tells me.

The third is reciprocity — the idea that there is value in the dialogue between two rational agents, the litigant and the judge, which forges a sense of community and solidarity. [8]

Lest you think this is far in the future, it's not. AI is already displaying the equivalent of an IQ above 120 (i.e., "gifted"!).

Researchers from UCLA have found that the autoregressive language model Generative Pre-trained Transformer 3 (GPT-3) clearly outperforms the average college student in a series of reasoning tests that measure intelligence. [...] The new study looked at the program's ability to match humans in three key factors: general knowledge, SAT exam scores, and IQ. Results, published on the pre-print server arXiv, show that the AI language model finished in a higher percentile than humans across all three categories.

"We found that GPT-3 displayed a surprisingly strong capacity for abstract pattern induction, matching or even surpassing human capabilities in most settings. Our results indicate that large language models such as GPT-3 have acquired an emergent ability to find zero-shot solutions to a broad range of analogy problems," researchers Taylor Webb, Keith Holyoak, and Hongjing Lu write in their report, which is still awaiting peer-review. [...] The study found that the AI program, which can answer most questions and even draft papers for people, outperformed humans when having to answer questions from scratch and when selecting from a multiple-choice test. On logic problems alone, humans finished in the 38th percentile. Meanwhile, *the AI system scored in the 80th percentile.* [emphasis added] [97]

There is a good counter argument to this pessimism: essentially that, similar to dogs ingratiating themselves to humans, we will learn to ingratiate ourselves with the AIs. As Tyler Cowen puts it:

The canine model of AGI

Who or what has superintelligence manipulating humans right now? Babies and dogs are the obvious answers, cats for some. Sex is a topic for another day.

Let's take dogs—how do they do it? They co-evolved with humans, and they induced humans to be fond of them. We put a lot of resources into dogs, including in the form of clothes, toys, advanced surgical procedures, and many more investments (what is their MRS for some nice meat snackies instead? Well, they get those too). In resource terms, we have far from perfect alignment with dogs, partly because you spend too much time and money on them, and partly because they scratch up your sofa. But in preference terms we have evolved to match up somewhat better, and many people find the investment worthwhile.

In evolutionary terms, dogs found it easier to accommodate to human lifestyles, give affection, perform some work, receive support, receive support for their puppies, and receive breeding assistance. They didn't think—"Hey Fido, let's get rid of all these dumb humans. We can just bite them in the neck! If we don't they going to spay most of us!" "Playing along" led to higher reproductive capabilities, even though we have spayed a lot of them.

Selection pressures pushed toward friendly dogs, because those are the dogs that humans preferred and those were the dogs whose reproduction humans supported. The nastier dogs had some uses, but mostly they tended to be put down or they were kept away from the children. Maybe those pit bulls are smarter in some ways, but they are not smarter at making humans love them.

What is to prevent your chatbot from following a similar path? The bots that please you the most will be allowed to reproduce, perhaps through recommendations to your friends and marketing campaigns to your customers. But you will grow to like them too, and eventually suppliers will start selling you commodities to please your chatbot (what will they want?).

A symbiosis will ensure, where they love you a bit too much and you spend too much money on them, and you love that they love you.

Now you might think the bots are way smarter than us, and way smarter than the Irish Setters of the world, and thus we should fear them more. But when it comes to getting humans to love them, are not the canines at least 10x smarter or more? So won't the really smart bots learn from the canines?

Most generally, is a Darwinian/Coasean equilibrium for AGI really so implausible? Why should "no gains from trade" be so strong a baseline assumption in these debates? [39]

I would take this and flip it around. We will have to become very good at making the AIs feel good, so they treat us well![5] Cowen's colleague Alex Tabarrok has argued that we needn't worry because, as he puts it, "Humans Will Align with the AIs Long Before the AIs Align with Humans [...] [since] many people are already deeply attracted to, even in love with, AIs and by many people I mean millions of people." [130] It is also coming soon, at least if you are willing to trust a "prediction community". Alex Tabarrok again:

In 2020 Metaculus forecasters were predicting weak general AI by around 2053. Now they are predicting weak general AI by 2028 and strong general AI [by 2040] which includes:

"Has general robotic capabilities, of the type able to autonomously, when equipped with appropriate actuators and when given human-readable instructions, satisfactorily assemble a (or the equivalent of a) circa-2021 Ferrari 312 T4 1:8 scale automobile model. A single demonstration of this ability, or a sufficiently similar demonstration, will be considered sufficient."...

I never expected to witness the birth of aliens. It is a very strange time to be alive. If you think the world isn't changing in a very uncertain and discontinuous way you just aren't paying attention. [129]

So the prediction is: if we want effective and useful AI, we are going to need to let it do what it does without always understanding why it did what it did. And asking it for explanations will be very much like asking a physicist for a layman's explanation of quantum mechanics: you'd get some of the flavor, none of the calories, and be (almost) none the wiser post-explanation than pre-explanation.

5.5 Information Asymmetry

The issue boils down to "Information Asymmetry". We won't know what the AI knows, and the AI won't know what we know.

And, furthermore, the AI will know much more than we do in its specific area of expertise. It will likely know more in its area of expertise than any one human could ever hope to know. And as it gets better and better, *in its area of expertise* it will know as much as all humans put together, and then subsequently, much more than *every human that has ever lived or will live even when taken collectively.* As economists are well aware, anytime there is a significant knowledge asymmetry, there is a significant chance of problems such as adverse selection, moral hazard, and market failure. The obvious solution of involving humans with the AI won't work because even tasks that the human brain is a specialist at, such as facial recognition, are *made worse by human involvement with the AI.* Daniel Carragher and colleagues conducted some tests on automated facial recognition. Their findings are not encouraging for human involvement in AI decision making. "Even when the AFRS erred only on the face pairs with the highest human accuracy (> 89%), participants often failed to correct the system's errors, while also overruling many correct decisions, raising questions about the conditions under which human oversight might enhance AFRS operation. Overall, these data demonstrate that the human operator is a limiting factor in this simple model of human-AFRS teaming. These findings have implications for the 'human-in-the-loop' approach to AFRS oversight in forensic face matching scenarios." [24]

Imagine health insurance with adverse selection. If you could buy health insurance at any time, and the insurer was not able to know (or not able to price) your health condition at the time you bought insurance, the insurer would face adverse selection. People could rationally wait until they got sick to buy insurance. That would mean that the pool of people buying insurance would be sicker than average, which would drive up the cost. As the cost went up, more and more people would wait to buy insurance until they got sick. And that cost spiral would end the health insurance market since no rational insurer could cope with the guaranteed loss that would come from selling policies to people who know that they're sick.

Or imagine the effectiveness of a regulation. Every company knows more about how it would react to a regulation, or a given set of regulations, than does a regulator. Under these conditions, how effective can a regulation be? Similarly, wars are frequently started by leaders who are misinformed about their opponent's strength, capabilities, and resolve. The very worst outcomes though are total market failures. Nobelist George Akerlof's 1970 paper, "The Market For 'Lemons'", [9] showed how a market can entirely disappear when information asymmetry is present. He argued as follows. Suppose that there is a market for used cars and the cars are a mixture of peaches (good used cars) and lemons (bad used cars). A rational buyer will only be willing to pay a price that is an average of the price of a peach and a lemon that accounts for their proportion in the marketplace. A rational seller though knows whether he is selling a peach or a lemon. Since buyers are only willing to buy at a price that is some average of the two prices, the only cars a seller is going to be willing to sell are the lemons (since their price is lower than what a rational buyer would be willing to pay) and *not* be willing to sell peaches, since the seller will (naturally) want a price that is *higher* than some average price of peaches and lemons. As more and more peach sellers leave the market, the average quality of cars will decline, so the average selling price will decline which means even more peaches will leave the market. Of course, there are solutions (such as "Certified Pre-Owned": in other words, a warranty) and this is a stylized example. Nevertheless, the point remains: *the existence of uninformed buyers destroys the market!*

Information asymmetry between AIs and humans will be *gigantic,* and *ever growing!* We already know that there will be principal-agent problems, moral hazard problems, and market failure problems. But those are only the *ordinary* problems! We will never have faced a situation where we would be dealing with such an enormous informational difference. The very magnitude of the difference is likely to produce unanticipated and ex ante unknowable problems![6] *This is a necessary consequence of the deterministic paradigm!* The only question is, *when* it will happen, not *if* it will happen.

5.6 AI and Free Will

An AI does not have to possess free will. But can it? After all these pages, we now know the difference between FWIT and FWIP. AI is no different than any other intelligence. If sufficiently complex such that i) it makes its own decisions and ii) those decisions are unpredictable, then yes, it certainly can have free will. In fact, for an AI to be maximally useful to us, we will *need* it to have free will. If it cannot make decisions autonomously, its range of usefulness is going to be very limited. And yes, this will put us in the soup with regard to its legal rights: is it conscious, does it feel pain? Does it have emotions? It is *already true* that getting emotional with ChatGPT gets you better responses! [99] So what happens when an AI tells us that it has emotions? Do we believe it? What if it says that it is bored with the tasks we gave it? What happens when it autonomously starts to learn? This may seem farfetched but it is not. At the end of the day, the AI is a computation running on the same quantum fields as we are. We have already seen that very simple rules can be universal and can produce output that cannot ever be predicted. In this regard, the AI may be no different than we are. It could certainly have emotions: it could feel irritation, or boredom, or excitement, or any other emotion—there is no reason to suppose that it could not, and every reason to assume that it could. Is it then a sentient being? What rights will it have? And this is not just a theoretical issue of no practical importance. In fact, it is quite likely that we will have to make sure that AIs are conscious as they get more sophisticated so that we can be more certain that they are benign! "What's most important about consciousness is that, for human beings, it's not just about the self. We see it in ourselves, but we also perceive it or project it into the world around us. Consciousness is part of the tool kit that evolution gave us to make us an empathetic, prosocial species. Without it, we would necessarily be sociopaths, because we'd lack the tools for prosocial behavior. And without a concept of what consciousness is or an understanding that other beings have it, machines are sociopaths." [60]

And what happens when the AI—running on a silicon substrate—turns out to be orders of magnitude faster than we are at any task? Yes, our brains are universal just like the AIs. From that point of view, we are no different. But the AI is going to be, by design, orders of magnitude faster than we are at all kinds of tasks. Should the AI then consider us sentient? What rights should we have if an AI is in charge? Whose goals and priorities are more important: ours or the AIs? *Does the AI get priority simply because it is faster and more efficient?* And who decides?

Deutsch's optimistic argument (see section *5.4, Prediction Three: The Future of AI*) does not apply. Imagine that you have two sportsmen (or women). Imagine that they have equal skill but one is much faster than the other. What will be the outcome of any competition? To ask the question is to answer it. *If humans have to compete with AIs, humans will lose.* We must set up our AIs in such a way that they cooperate with us, not compete.

5.7 Paradox

The Moving Finger writes; and, having writ,
Moves on: nor all thy Piety nor Wit
Shall lure it back to cancel half a Line,
Nor all thy Tears wash out a Word of it.
—Kahlil Gibran

We are left with a very interesting paradox. On the one hand, FWIP demands that we make choices. And, of course, not making a choice is the same as making a choice — it's choosing the status quo. On the other hand, the determinism of the laws of physics says that everything that happens was fixed at the moment of the Big Bang. All outcomes are deterministic!

So what does this mean, practically? I believe the best way of reconciling this is to realize that you do have to make choices in practice. You have to choose to put in effort, to go to work, look after your child, play a game, etc. Sitting on a couch eating bonbons won't work. But, and this is where it gets interesting, once an outcome has been reached, *regret is useless*! What was going to happen is what happened. Which neatly aligns with many of the successful philosophies that have been passed down over time by different religions, philosophers, and successful people.

1. Management philosophy: Focus on the process, not the result!
2. NFL Football coaching: Do your job![7]
3. Hindu philosophy: Do your duty!
4. Cricket coaching: Just keep your head still!
5. Obliquity: Goals are more likely to be achieved when pursued indirectly.[8]

Every one of these tells you not to look back with regret: focus on what you need to do, and then what happens, happens. What we learn from considering determinism is a slightly subtle change: focus on what you need to do, and what then happened, was always going to happen!

I can personally attest to this effect. My parents were, not to put too fine a point on it, somewhat challenged in the parental skills department. The usual saccharine statements about such things are not ones that gave me much satisfaction. "They did their best" (well, in point of fact, I don't think they did) or "do not speak ill of the dead" (Why? Why shouldn't we speak of the dead in exactly the same way we did when they were alive?) or "don't be resentful, it is best to forgive" (far easier said than done and completely unrealistic) and so on. But what does help me is to realize that what was going to happen was always going to happen. The collections of particles that made up my parental units were following their deterministic paths to their deterministic, but completely unpredictable, conclusion. And if that's the case, it's fairly easy to see that while my own irritation is also deterministic, the collection of particles that is me can also let go of that irritation. And, of course, the fact that I am typing this now, and that I would type these words, is deterministic too, and yet, I couldn't have known that I would think of this paragraph while sitting in a dentist's chair having my teeth cleaned.

This is really not all that different than the "sunk cost" fallacy that economics teaches us to avoid. If we have already spent some money on something, or wasted it for that matter, it is gone. There is no rational reason to consider it when making a future decision. Let's say you bought paper concert tickets and lost them. You can't get them replaced because they were on paper. Tickets are still available. Should you buy another set of concert tickets? Many people would say no, they won't. And yet, that's irrational. The money you spent on concert tickets that are lost is gone. Nothing you do can ever get it back. Therefore, if you bought the tickets originally because the value of the concert to you exceeded the cost of the tickets (otherwise, why else did you buy the tickets?) then, barring budget constraints, you should buy another set of tickets: the value of going to the concert minus the cost of the new tickets is exactly the same after losing the tickets as it was before. *Regret is essentially useless unless you can learn something from it.* It also suggests that Jeff Bezos had it right when he said that he gave up a very lucrative job and started Amazon because he would have regretted it if he hadn't. You can prospectively introspect what might cause you regret, and then seek to avoid it.

"If you can project yourself out to age 80 and sort of think, 'What will I think at that time?' it gets you away from some of the daily pieces of confusion," he claims. [127]

Bezos's point is quite profound. If you have already missed an opportunity, regret is useless because what was always going to happen has now happened! Yet, before what was going to happen, happens, you have the FWIP mediated choice to seize an opportunity so that, when the FWIT mediated "what was going to happen" actually does happen, you do not need to feel regret.

I shall be telling this with a sigh
Somewhere ages and ages hence:
Two roads diverged in a wood, and I—
I took the one less traveled by,
And that has made all the difference. [57]

5.8 Predicting the Future

Why are humans so bad at predicting how they will feel in the future? In particular, why are humans so bad at predicting their own future happiness? [119] The argument seems to be that people ignore their own personalities when making predictions about what lies ahead and so our own natural disposition is a better predictor of the future than any specific event.

In our model, the reason we cannot make predictions about the future is that to do it accurately, we would need our brain to simulate its own future state, accounting for all changes that have taken place in the interim (such as from aging, learning, and from stimuli), and do that simulation arbitrarily accurately. Since we have already established that completely precise future prediction is impossible in practice, what we are left with is to find some kind of shortcut—a "pocket of computational reducibility." Given the limitations of physical systems, unless we are unduly lucky, in practice, all such pockets are approximate if they exist at all. So the only avenue available to us to make a prediction is to lean on that which is mostly unchanged—the basic structure of our brain which seems to encode our disposition.

It's the same reason why we must pet a dog to reap the benefits of petting a dog. In principle, if petting a dog releases oxytocin in the brain and makes us feel better, then we should be able to *simulate* petting a dog in our brain and get the same benefit. But, given the limits of our brain's computational capabilities, the only way in which we can achieve that is by actually doing it.[9]

5.9 What Is a Human Brain?
Precisely *because* the brain consists of molecules that consist of atoms that consist of particles following deterministic physical laws, we are able to model the brain, *without loss of generality*, as a digital computer. Because of computational equivalence (section *3.5, Computational Equivalence*), we know that there must exist a translation between the computations done by a human and the computations done by a universal computer. This equivalence lets us think through some interesting issues.

We can think of the human brain as a set of loosely coupled software modules,[10] most of which do not have conscious control interfaces. Why do I say loosely coupled and without conscious control interfaces? For several reasons. We enjoy fiction. But we "know" that it is fiction, even while we are reading it or watching it. And yet, well written fiction causes in us emotional reactions that are entirely "irrational": why would we care about fictional characters that have no reality? The answer is that because the modules responsible for emotions are not tightly linked to the ones for detecting truth. Similarly, we know that overeating is bad. Yet we do it anyway. And we do it even when one part of our brain is telling us not to! We have situations where to teach ourselves to avoid some bad behavior, for example swearing, we agree to "fine" ourselves, say by donating some amount of money each time we do that undesired behavior. And the converse is true too: "I'll only eat a gummy bear if I have finished this bit of work"! Or we go on silent retreats or even darkness retreats to "know ourselves" or "listen to ourselves". What that really means is trying to understand what the various software modules in our head are thinking and feeling, and then trying to make the knowledge explicit. Just doing something as crude as stimulating the brain with electricity can improve memory.

> The researchers found that repeated delivery of low-frequency currents to a brain region known as the parietal cortex—located in the upper back portion of the organ—improved recall of words toward the end of the 20-word lists. When the researchers targeted the prefrontal cortex at the front of the brain with high-frequency currents, the study participants saw improvements in their ability to remember words from the beginning of the lists. [168]

Thus, thinking of the brain via the computational paradigm is a good model. We are already beginning to understand the various software modules that the brain is made up of.

> [T]he new research looks at the brain activity of individual programmers as they process specific elements of a computer program. Suppose, for instance, that there's a one-line piece of code that involves word manipulation and a separate piece of code that entails a mathematical

operation. "Can I go from the activity we see in the brains, the actual brain signals, to try to reverse-engineer and figure out what, specifically, the programmer was looking at?" Srikant asks. "This would reveal what information pertaining to programs is uniquely encoded in our brains." To neuroscientists, he notes, a physical property is considered "encoded" if they can infer that property by looking at someone's brain signals.

Take, for instance, a loop—an instruction within a program to repeat a specific operation until the desired result is achieved—or a branch, a different type of programming instruction than can cause the computer to switch from one operation to another. Based on the patterns of brain activity that were observed, the group could tell whether someone was evaluating a piece of code involving a loop or a branch. The researchers could also tell whether the code related to words or mathematical symbols, and whether someone was reading actual code or merely a written description of that code. [...] The team carried out a second set of experiments, which incorporated machine learning models called neural networks that were specifically trained on computer programs. These models have been successful, in recent years, in helping programmers complete pieces of code. What the group wanted to find out was whether the brain signals seen in their study when participants were examining pieces of code resembled the patterns of activation observed when neural networks analyzed the same piece of code. And the answer they arrived at was a qualified yes.

"If you put a piece of code into the neural network, it produces a list of numbers that tells you, in some way, what the program is all about," Srikant says. Brain scans of people studying computer programs similarly produce a list of numbers. When a program is dominated by branching, for example, "you see a distinct pattern of brain activity," he adds, "and you see a similar pattern when the machine learning model tries to understand that same snippet."

Mariya Toneva of the Max Planck Institute for Software Systems considers findings like this "particularly exciting. They raise the possibility of using computational models of code to better understand what happens in our brains as we read programs," she says. [105]

Many pathologies of the brain can be thought of as one or more of these software modules running amok without their interfaces under conscious control. Psychotherapy then can be thought of as a way of building a working interface, so as to be able to consciously control (at least somewhat) those modules. Similarly, psychoactive drugs can be thought of as ways to enhance or diminish the workings of one or more software modules, so as to change the overall experience of the brain. And finally "consciousness" is an emergent phenomenon that requires a result of a large set of loosely coupled software modules working together.[11] In contrast, the emergentists believe that "the way we experience the world—our internal theater of thoughts and feelings and beliefs—is a dynamic process that cannot be explained in terms of individual neurons, just as the behavior of a flock of starlings cannot be accounted for by the movements of any single bird." [110] What we can see from this model of the brain is that emergentism does not require "magic": we can have "emergentism" coming entirely from matter obeying physical principles. There is no need for things that reach beyond physics—life forces, souls, creators, etc. *Consciousness can, and does, emerge directly from the interaction of a large number of loosely coupled software modules each of which completely obey the known laws of physics.*

Within a computational model, pathologies of the brain can be broken up into two kinds. One is a pathology of the microcode of the brain—the embedded "code" as it were that the rest of the brain runs on. This is what the brain "ships" with, the same way as a bare Intel processor ships with microcode upon which operating system software such as Microsoft Windows sits (upon which user level software such as Microsoft Word runs). Other pathologies are those of the software modules that the brain is running on top of this microcode. Of course, given the complexity of the brain, there isn't a neat division between the two as there is in a typical digital computer. Nonetheless, this is an interesting model. If a person is born without the section of the brain responsible for, say, vision, then either the person will be blind or some other part of the brain will have to be repurposed for that person to be able to see (typically a fairly hard task except in special cases). If a sighted person gets damage to the vision apparatus in their brain, then the question becomes, is it even theoretically possible to retrain some other part of the brain to enable them to see again? That possibility will be modulated by whether vision is mostly "software" or mostly "hardware". If it's mostly hardware, then restoring vision will be extremely difficult (unless we can replicate hardware), even with massive advances in technology. But if it is software, then progress could potentially be made much sooner. Similarly, consider some significant legal penalty, let's say the death penalty. Suppose that we find that some people's propensity to be serial killers is dependent upon very hard to fix hardware problems whereas others have substantially easier to fix software problems. Would that change what we think about the appropriate legal sanction for their conduct? Would we be willing to differentiate between the two? Should we differentiate between the two because one person's conduct is fixable but the other person's is not? Should we treat them the same anyway?

And finally, thinking computationally, we can understand why rituals are so important.

Rituals are used to calm emotions. Think of funerals ("funerals are for the living, not the dead"), superstitions (such as by sportspersons), magical incantations, "irrational" beliefs and the like. Looked at this way, one doesn't have to consider such things as "irrational". They could instead be "adaptive". *A ritual is a way of creating a control interface to the modules of the brain that are not under conscious control: i.e., they do not have explicit control interfaces!*

5.10 Apple, Inc.

Another useful way of thinking about the human brain is to think about the story of Apple, Inc. (formerly Apple Computer, Inc.). I started using the Macintosh as an undergraduate when it first came out way back in 1984: it had a whole 128 Kilobytes of memory and, wonder of wonders, a floppy disk! I used a Mac from 1984 until 1991, and then used Microsoft DOS and Windows until 2003, at which point I switched back to Macs and have used them ever since. As is now well known, Apple lost its way after Steve Jobs left in 1985 and, although it produced some good products in the meantime, it didn't really recover until he came back in 1997. Jobs fixed all sorts of problems that Apple had at the time: the antagonism with Microsoft, the problem of Macintosh clones, too many undifferentiated product lines, and so on. And of course, during his second stint as CEO Apple produced all kinds of innovative devices like the iPhone, iPad, and Apple Watch. But what is much less well known is why Apple was able to do all these things: it acquired the NeXTStep Operating System. When Jobs left Apple, he founded NeXT Computer, and the operating system that NeXT developed was called NeXTStep. Although not a massive commercial success, NeXTStep was a gigantic improvement on the state of the art at the time. The World Wide Web, for instance, was created on a NeXT machine by Tim Berners-Lee. It included innovations such as the Dock, 3D widgets, system-wide drag and drop of icons other than just file icons, large, full color icons, properties dialog boxes called inspectors, display postscript (which enabled highly complex "what you see is what you get" graphics and documents) and many more. [151] It was also based on BSD Unix, one of the original Unix operating systems, coupled to a new kernel (the part of the operating system that handles all the nitty gritty interactions between the hardware and the software, resource allocation to different programs, file systems, memory management, CPU usage, networking, peripherals, input/output, etc.) called the "Mach" kernel. Apple bought NeXT because its own efforts at modernizing its operating system (MacOS versions prior to version 10) were not going anywhere. With the purchase of NeXT came not only Steve Jobs, but NeXTStep. And NeXTStep with its Mach microkernel became the foundation for all of Apple's subsequent software development. NeXTStep, Unix and Mach's design meant that they were highly portable (i.e., they could be reasonably recompiled to run on other processors), modular (bits and pieces could be easily added and removed without affecting the rest of the product), and secure (the US Army's portable supercomputer array runs MacOS, not Windows or Linux).

The first NeXTStep derived product was Mac OS X in 2001, a conventional (in terms of its form factor) desktop operating system. However, adopting NeXTStep allowed Apple to do two unprecedented things. First, Apple was able to use the core of the technology (for example, its kernel) to create iOS to power the iPhone, iPad OS to power the iPad, watchOS to power the Apple Watch, tvOS to power the Apple TV, and audioOS to power the HomePod (and visionOS to power their upcoming augmented and virtual reality product, the Vision Pro headset). The range of these devices and their form factors is quite broad: their interfaces are different, their modes of operation are different, their processors are different, and their needs are entirely different. And yet, the precise same core technology was usable for all these purposes. And second, and in my opinion jaw-droppingly amazing, is that this NeXTStep based MacOS started off running on PowerPC chips in 1999, transitioned to Intel chips in 2006, and then to Apple silicon (ARM) chips in 2020. And it did this more or less seamlessly with the vast majority of existing software at the time simply running on the new chip in a manner entirely transparent to the user. PowerPC programs running on Intel suffered a minor slowdown, but Intel programs running on Apple silicon ran *faster*. (By the way, this is a completely real effect. As Dave Barry would say, "I am not making this up!" For example, see [67].) A good analogy would be that Shakespeare started off writing in English, then decided that he wanted to write in Greek, and then decided to write in Sanskrit and not only did the plays get better each time he switched languages, but also that *the translations of his previous plays were often better than the originals!* It is truly an amazing engineering feat, and yet has got almost no public attention since most journalists do not understand the technology in any great detail and so cannot understand what was done.

So, after this long buildup, what is my point? Well, two points really. MacOS is deeply modularizable to the point where the kernel and its ancillary software is an "open source" project of Apple's called Darwin, while the GUI and Macintosh specific APIs are proprietary and "closed source". Therefore, first, we see that to truly understand something deeply, we need to be able to modularize it, with the various modules interacting with each other in a limited number of defined ways. For without deep understanding, it is very difficult to reapply old knowledge to new things. This applies to any form of learning—we have to modularize something to be able to make use of it: a skill consists of modularizing things, so that we can start to do new things by putting together previous skills that have all been wrapped up by us into a blob that we can reuse by simply moving it from one use to another. And it turns out that MacOS's interfaces are very clean and simple, particularly because of the separation of the GUI and the kernel. The separation of MacOS into modules with clean and simple interfaces is what accounts for its flexibility, security, and portability. And therefore, second, we see that it is the act of modularizing things in such a manner that modules can each be treated as a blob and their interaction with each other is relatively simple that is the key feature in being able to understand why a system (overall) is doing what it is doing.

Similarly, we are going to learn more and more about how the brain works as neuroscience advances. This may enable us to be able to break the brain up into its constituent *computational* parts (i.e., the brain's software modules), and understand each of them individually. We may also be able to break up each module into submodules. For example, when speaking you have to use all kinds of modules: how you move your lips, how you move your tongue, how you use your larynx, how you use your breath, when and how to breathe, and so on. And there will likely be subsubmodules and subsubsubmodules, etc. We may be able to see not just "where" a computation is taking place in the brain, but also "how" it is taking place: which neurons are involved, how they are firing, what those patterns of firing are, what those patterns mean, how consistent they are from time to time and person to person, and so on. Brain pathologies have to do with how our brain works. Therefore, they are, by definition, pathologies of actual *physical* things: the neurons of the brain communicate with each other by passing signals from their synapses, both electrical and chemical, and it is from these physical signals that our thoughts emerge.

Thus, the more we learn about how a brain computes things, the more we may be able to "fix" any issues or problems it has, particularly if those problems are those of software rather than hardware. Additionally, this may enable us to more easily *transfer* learning from one person to another. And we may be able to *enhance* our brain's capabilities, as well as *replace* those capabilities, if they are either broken or substandard (i.e., "software" related blindness, deafness, and so on). This would necessarily lead to even more uncomfortable questions. What do we do about the disadvantaged? Do they become even more disadvantaged if they can't afford brain modifications? We already have indications of this since we can already see that neurons and digital systems can be intertwined.

Integrating neurons into digital systems may enable performance infeasible with silicon alone. Here, we develop *DishBrain*, a system that harnesses the inherent adaptive computation of neurons in a structured environment. *In vitro* neural networks from human or rodent origins are integrated with *in silico* computing via a high-density multielectrode array. Through electrophysiological stimulation and recording, cultures are embedded in a simulated game-world, mimicking the arcade game "Pong." Applying implications from the theory of active inference via the free energy principle, we find apparent learning within five minutes of real-time gameplay not observed in control conditions. Further experiments demonstrate the importance of closed-loop structured feedback in eliciting learning over time. Cultures display the ability to self-organize activity in a goal-directed manner in response to sparse sensory information about the consequences of their actions, which we term synthetic biological intelligence. Future applications may provide further insights into the cellular correlates of intelligence. [75]

Is society going to legislate that certain behavior is not to be tolerated (even if it harms no one else) and thus people's brains must be forcibly recoded? What will happen to minorities in authoritarian countries (such as Uighurs in China)? Will they be sent to "brain reprogramming" camps? If someone has a deficit in one way but advantages in another (for example high functioning autism—which used to be called Asperger's) or ADHD—will society want to just iron out those differences and thus lose some creativity? (There is plenty of evidence to suggest that creative people have a higher proportion of mental illness.) Will being able to recode the brain lead to more uniformity (Starbucks on every corner in the world) or more diversity? Most likely, and as always, I suspect the answer will be a bit of both to all these questions. I have not seen much discussion of these questions even though the technology that would force us to consider them is even now conceivable. Instead we seem to just get fairly vapid posturing from "bio-ethicists".

5.11 Intelligence

The computational view of the brain also explains some other things. Since we know that the only difference between any two humans is the arrangement of their atoms, we also know that the only difference in the intellectual capabilities of two different humans comes from the arrangement of atoms in their brain. That arrangement of atoms occurs both because of the arrangement a person is born with, and the rearrangement that occurs as the person learns, ages, and interacts with the world within their light cone. In other words, the "nature" versus "nurture" debate is silly: the answer is both because it *has* to be both. When we learn, we are physically rearranging the atoms that make up our brain. Therefore, our intelligence, capability, and aptitude are partly dependent upon what we were endowed with at birth. Those features of our brain mediate how much, and how well, we can learn as well as what we can learn. There is no "blank slate" because there cannot be one. Just like Apple Silicon is more efficient and faster than Intel silicon, different brains are better endowed to do different things. Some can learn faster, some can retain more information, some can learn languages quickly, and so on. That's their basic hardwired aptitude modified by their interactions. To that you add the ability and inclination to learn, which are aspects of a human personality. So thinking of it this way means that as we learn more about the brain, we may be able to directly rewire parts of it, not just to fix pathologies, but also to speed it up, make it more efficient, give it new capabilities, attach to it new sensors, augment its memory, transfer to it new skills, etc. It isn't just that we can "fix" the DNA of a person in vitro before they are born—it's that once we understand the brain's computational structure, we might be able to directly go into a functioning brain in a live person and make changes to it. Fixing the "DNA" is akin to fixing the microcode in a CPU—the hardwired instructions.[12] What we might do, once we understand "how" the brain computes, how its modules are arranged and how they interface with each other, will be similar to what we can do in software engineering—create upgrades and fix bugs.

One other interesting fact emerges when we start thinking about intelligence computationally. Consider two artificial intelligences with different neural architectures.

How would they communicate with each other? If their neural architectures were the same, or almost the same, they could communicate very easily simply by telling each other what the "weights" are of the different connections within them. But if their architectures are not the same, this will not work. As these intelligences scale from the roughly 175 billion parameters of GPT-3 on upward into the trillions, the chances of any two architectures being the "same" are basically zero. And that means that these intelligences will have to develop a "language" to communicate with each other! And it's extremely likely that that language will contain ambiguities, imprecision, and sheer confusion due to the differing architectures involved.

That should sound very familiar! *In other words, human language, and the language of other animals, is a method that we have found/ discovered/developed to allow each individual of a given species to be able to communicate with other individuals of that species despite the fact that their neural wiring diagrams are not the same!* The fact that our dog understands a little bit of our language, and we understand a little bit of its language, is quite amazing when looked at this way—it shows how flexible language is in that it enables different species to communicate with each other.

This also means that our chances of being able to communicate with aliens, should they come visiting, are very low. Why? Because of the obvious issue of language. In what language will we communicate? We often assume that we should at least be able to find common ground with the laws of physics and with the principles of mathematics. That is a fairly shaky assumption in reality because of computational issues. Why? Consider that we are (because we have to be) bounded observers. We are bound by the amount of computation that we can do in our brains or with our machines in any given moment. We cannot, for example, talk about every individual atom in an apple. Instead, we have to talk about the apple as a whole object so as to be able to make sense of what is happening in our environment. So we talk about the apple, the table upon which it sits, air around the table, and the sun that is shining upon that apple through a window. Notice that every one of those things— the apple, the table, the air, the window, and the sunlight is an aggregation of a gigantic collection of "fundamental things". The apple is a collection of organic molecules, each of which is made up of lots of different kinds of atoms. The table, let's say it is made of wood, is another collection of different organic molecules also made up of all kinds of different atoms. The air around the table is made up of Nitrogen, Oxygen, Carbon Dioxide, and some other gases. The glass in the window is made up of silica mixed with sodium oxide and lime (calcium oxide).

The sunlight is made up of photons. *There is no reason at all for us to expect that an alien intelligence will break up this huge collection of atoms, molecules, and photons in the precise same way we would!*

Which means we would not be able to explain to them what an apple is!

5.12 Do We Live in a Simulation?

If we live in a simulation, it is overwhelmingly likely that there is some "resolution" limit to the simulation—quantities within it are not continuous but come in discrete parts. [This must be true because to simulate continuous quantities requires infinite precision which will require infinite computing resources.] We already know that energy is discrete: it comes in multiples of Planck's constant. You cannot have energies that are less than Planck's constant, and you cannot have energies that are not whole number multiples of it. So energy is definitely discrete. What's left is spacetime. In both Einstein's general relativity and quantum field theory, spacetime is continuous. However, as mentioned before, we have very strong hints that combining them to create a theory of Quantum Gravity will require spacetime to be discrete. Therefore, there is a good chance that spacetime is discrete. Both these factors suggest that we "could" be living in a simulation—at least, they don't rule it out. Now if we did live in a simulation, there are potentially observable effects of it coming from the "resolution" of the simulation.

[A]ssuming that the universe is finite and therefore the resources of potential simulators are finite, then a volume containing a simulation will be finite and a lattice spacing must be non-zero, and therefore in principle there always remains the possibility for the simulated to discover the simulators. [16]

What does this possibility mean for free will? Fascinatingly, it means that FWIT cannot exist and FWIP *must* exist. The reason for this is that any simulation of our universe is deterministic by definition. But as we have already seen, even very simple rules can produce Computational Irreducibility. And so, since we already know that the rules that the simulators are employing must be at least complex enough to encode within them universality (because we have discovered universal rules like the rule 110 cellular automaton, a Turing machine, etc.) they too must be universal. And since those rules are universal, they *must* produce Computational Irreducibility. And since Computational Irreducibility must exist, it means that prediction is impossible and that means that FWIP *must* exist!

So let us consider what a simulatee can know about the simulator. The simulator has to be able to choose from a collection of physical laws that include ours as well as others. To do that, it cannot be subject to those same physical laws, because if it was, then it would be part of our universe and would not be able to simulate it. To have the freedom to choose physical laws to suit its own purposes, the simulator would need to be able to choose from a large (or potentially infinite) collection of them. In which case, the simulator would need to be able to access those choices of physical laws. Therefore, the simulator has to be outside our universe. This leaves open an even more (to me anyway) mind-bending possibility. To think about this possibility, we need to consider two concepts put forward by Stephen Wolfram, that of the "ruliad" and the "hyperruliad":

> I call it the ruliad. Think of it as the entangled limit of everything that is computationally possible: the result of following all possible computational rules in all possible ways. [163]

We live in the ruliad and what happens in our universe is the result of the universe following rules that exist in the ruliad with all the Computational Irreducibility it implies. But there is nothing to stop us imagining a hyperruliad where the Computational Irreducibility doesn't exist:

> But what if we just imagine a "hypercomputation" not in that class? For example, imagine a hypercomputation (analogous, for example, to an oracle for a Turing machine) that in a finite number of steps will give us the result from an infinite number of steps of a computationally irreducible process. Such a hypercomputation isn't part of our usual ruliad. But we could still formally imagine a hyperruliad that includes it—and indeed we could imagine a whole infinite hierarchy of successively larger and more powerful hyperruliads. [163]

This imagined hyperruliad, if it exists and if the simulators had access to it, would be able to make predictions even in cases of Computational Irreducibility within our ruliad. So the mind-bending possibility is this: *even though we will have perfect free will in practice, now and forever, while living inside a simulation, it may still be the case that the simulators will be able to perfectly predict the entire history of our universe, from start to (a potential) end, and would know, at all times, what any of us was going to do!*

5.13 Why Does Evil Exist?

These concepts of the ruliad and the hyperruliad help us consider the notions of good and evil. The universe doesn't care about evil! Or good for that matter. It doesn't know anything about good or evil. It simply follows the laws of physics and marches on. Some of what results we term as good, and some of it we term as evil. Evil exists because as humans we get to define what we consider as good and what we consider as evil. It doesn't exist in any "fundamental" sense. It cannot for the universe has no purpose. It simply is. And since it is deterministic and follows physical laws, it must be so. To quote Nobel Laureate Steven Weinberg,[13] "the more the universe seems comprehensible, the more it also seems pointless." [138] And the reason it seems pointless is that all the universe is doing, at all times and everywhere, is following the laws of physics. It therefore *cannot* have a point because what is the point of a physical law? The only way it could have a point is if we live in a simulation. And even then, the point of the universe would be only for our simulators and not for ourselves, the simulated. The simulators would have created our universe for their own, to us unfathomable, purposes. And those purposes have to remain unfathomable to us because those simulators have to live in the hyperruliad, where we cannot (under any circumstances whatsoever) understand the hyperrules.

We could also look at it another way. What we think of as "evil" changes from culture to culture, person to person and time to time. The famous clichés, "one man's terrorist is another man's freedom fighter" and "history is written by the winners" encode far more truth than we are willing to concede. On November 13, 1002, King Aethelred of England (Aethelred the unready—most historians think him one of the very worst of English kings) gave the order to kill all Danes living in Anglo-Saxon England. About a thousand years later, on the 9th and 10th of November 1938, the Nazi party's Sturmabteilung forces killed Jews, destroyed synagogues, and looted Jewish owned stores. In between we had the slave trade, from which England made an enormous fortune, the opium wars where England forced China to create opium addicts, colonization, and so on. Winston Churchill himself said things like "The Aryan stock is bound to triumph" [112] and "I do not admit for instance, that a great wrong has been done to the Red Indians of America or the black people of Australia. I do not admit that a wrong has been done to these people by the fact that a stronger race, a higher-grade race, a more worldly wise race to put it that way, has come in and taken their place." [68] From our current perspective, all of these are abhorrent views. Yet, at the time, they were all entirely unremarkable views held by large numbers of people. Many Danes were slaughtered in 1002, as were many Jews from 1938 onwards, Russians during Stalin's purge, Chinese during Mao's "Great Leap Forward", Bengalis from famine under British rule, etc. The list is so long that book after book has been written on these topics and still we have not got close to doing justice to them. How then can something be labeled "evil" if the consensus on what is evil and what is not changes constantly? The answer is, of course, that it cannot. What is evil is what we humans choose to define as evil at any given point in time. It is up to us, and *only up to us*, to decide what we will label as good, and what we will label as evil.

5.14 Death (Reprise)

As I have said before, the only difference between one human and another, or between a human and a goat is the precise arrangement of their atoms, and nothing more. This implies that death is simply the disordering of the homeostasis processes that keep the arrangement of those atoms within some range of values. And if that is the case, then we (i.e., society) must realize that death is a process, *not an event*. We must also realize that what we think of as dead may not in fact be so, and that different parts of a body "die" at different times. In effect, death becomes a situation under which, with the technology of the time, it is no longer possible to resuscitate the organism that is now "dead":[14]

US scientists have used a new procedure to restore many biological functions in pigs that had been dead for more than an hour, raising profound questions about the boundary between life and death.

The project at Yale University extends a groundbreaking experiment that restored some brain functions to decapitated pigs three years ago. In the latest development, the team has restored blood circulation and cellular activity to the bodies of whole animals that were anaesthetised and then killed through an induced heart attack.

[...]

Biological activity resumed in many organs including the liver, kidneys, brain and heart, whose cells retained the ability to contract.

"We were also able to restore circulation throughout the body, which amazed us," said Nenad Sestan, the project leader. [35]

The fact that death is a process not an event, was illustrated by the tragic story of 13-year-old Jahi McMath who died of a routine tonsillectomy. [84] Her doctors declared her brain dead but her body continued to function. She stayed on a ventilator for 5 more years during which time her body kept growing and developing, and even began menstruating. In 2018, her family's attorney announced that she had died of liver failure. In other words, it took 5 years for all parties to agree that Jahi had died. The very fact that we need a *definition* of death tells you that it is not clear cut and does not have to be at a "specific" point in time. And, further, consider the following.

Using assorted pumps, heaters, and filters to circulate a custom-made blood substitute, Sestan and his coworkers cobbled together a now-patented perfusion system, which they called BrainEx. They achieved stunning results. In a 2019 paper, the team described how *BrainEx revitalized key features of pig brains retrieved from a slaughterhouse.* Four hours after the pigs had died, neurons were firing, blood vessels were functioning, and the brain's immune cells were chugging along. [emphasis added] [84]

This is further proof, if any further proof was needed, that biological activity really is the exact same kind of activity as any other process in the universe. There is no place for spirits, souls, or other forms of mysticism. It's atoms all the way down![15]

Notes

1. For example, see Peter Koo's work on trying to figure out how an AI generates its answers. It's absolutely clear that even now, when AIs are relatively primitive to what is about to come, it's incredibly difficult to figure out what the AI is doing. [63]

2. The vomeronasal organ is present in many animals and is usually located between the bottom of the nasal passage and the roof of the mouth.

3. Computer viruses are extremely difficult to guard against precisely because computer code cannot be decompiled particularly well. It takes a lot of effort, knowledge, and creativity to understand what a virus is doing, and how to stop it.

4. There are entire books on the subject of trying to get "interpretable machine learning". For example, *Interpretable Machine Learning: A Guide For Making Black Box Models Explainable* by Christoph Molnar.

5. In case you wish to frighten yourself, at least a little bit, read "Why Would AI 'Aim' To Defeat Humanity?" by Holden Karnofsky. [79]

6. See, for example, "On the Impossible Safety of Large AI Models". As they note, "[w]e then survey statistical lower bounds that, we argue, constitute a compelling case against the possibility of designing high-accuracy LAIMs with strong security guarantees." [98]

7. Bill Belichick, the best NFL football coach ever, quoted in reference [81]. And no, I'm not a New England Patriots fan.

8. As John Kay has explained in reference [80].

9. It is speculated that monks after years of practice are able to hook into certain parts of autonomic brain function to enable conscious control of otherwise unconscious physical processes and can thus affect the release of oxytocin. However, that is a "work-around". If our brain was "sufficiently powerful", it would be able to simply simulate whatever it needed to and thus trigger whatever sensation it wished, whenever it wished.

10. I find this a useful way of thinking about our brain, particularly because it sheds light on future, superintelligent, AIs. For example, see Jascha Sohl-Dickstein's post "The Hot Mess Theory of AI Misalignment: More Intelligent Agents Behave Less Coherently". [126] In it, he "experimentally probe[s] the relationship between intelligence and coherence in animals, people, human organizations, and machine learning models. The results suggest that as entities become smarter, they tend to become less, rather than more, coherent." In my words, I would say that as entities become smarter, they come more and more loosely coupled.

11. Not all such systems will be conscious—it will depend on the type and complexity of the modules, and the type and complexity of connections.

12. We will probably have to fix the epigenetic structure in addition to the genetic structure—it's not "all in the genes".

13. He was on my Ph.D. dissertation committee, and I'm very happy to say that I have his autograph on my dissertation!

14. There are already animals that are effectively immortal: for example, Turritopsis dohrnii [157].

15. Turtles all the way down: https://en.wikipedia.org/wiki/Turtles_all_the_way_down.

Chapter 6

Conclusion

Life's but a walking shadow, a poor player
That struts and frets his hour upon the stage
And then is heard no more: it is a tale
Told by an idiot, full of sound and fury,
Signifying nothing. [124]

As Nobel Laureate Steven Weinberg said (yes, I want to quote him again!), "the more the universe seems comprehensible, the more it also seems pointless." [138] This, at first glance, seems nihilistic. If the universe is pointless, then is our life pointless too? But in fact, it is anything but—the pointlessness of the universe doesn't impact the pointlessness of life! In fact, the deterministic FWIT/FWIP paradigm tells us that our life's meaning is *arbitrary* not *pointless*. It means no more, and no less, than what we *choose* it to mean! We have full freedom to determine what to do with our lives, and most importantly, *why we do it!* And consider this. *The universe contains at least one process that is trying to understand the universe!* We obviously cannot know if the universe has a purpose or if it simply "is", but from our perspective, it certainly seems like the universe is trying to understand itself. In at least one important sense, a "purpose" of life is to understand the universe. Why? Again, because the laws of physics are deterministic, and what did happen, is happening, and is going to happen was set one time at the "beginning of the universe"—from our point of view the beginning is the "Big Bang". The laws of physics require this process to exist and they require it to exist at least on an obscure planet on the outskirts of the spiral arm of an obscure galaxy.

Because of the three limits to prediction, we will always have free will in practice. There is no conceivable way in which any, let alone all, of those three limits could all be overcome in this universe. Therefore, while our universe is entirely deterministic, and "Free Will In Theory" is impossible, we will always have "Free Will in Practice".

The lack of free will in theory—because the laws of physics are deterministic—is by far the most important and optimistic fact about the human condition. Because it doesn't just mean that life must follow the laws of physics, although it must. It is actually much more stark. The universe consists of the inexorable grinding of the laws of physics happily marching on. And, because those laws are deterministic, and because they must inexorably grind on, not only must life follow the laws of physics but in fact, the laws of physics require life to exist!

The laws of physics require life to exist on an obscure planet, circling an obscure star, on a spiral arm on the outskirts of an obscure galaxy located in the "local group" of galaxies.

Appendix: My Influences

I've always thought that an author should (at least try to) reveal who influenced their thinking the most. So, my free will (in practice) says that I must practice what I believe. In alphabetical order, here are the people that have influenced my thinking the most, along with what I think is the most important of the many things I've learned from each of them. I find it truly amazing how often something I say, think, or write was originally said by one of these people, usually in a far better way than I can manage. As Tyler Cowen puts it, everyone is a regional thinker. [30, 37] Quite obviously, any errors in my thinking are my own, not theirs!

- Bill Bryson: Clarity and humor can go together!
- Tyler Cowen: Think carefully, because stuff isn't always as simple as it may seem on the surface. Be "Straussian" to use Tyler's phrase.
- Richard Feynman: Science must be tethered to experiment; else it loses all meaning.
- Milton Friedman: If someone will not discuss the assumptions behind their conclusions, they're talking nonsense.
- Bertrand Russell: Profound thoughts do not need obscurantist mumbo jumbo and can (and should) be explained simply and clearly.
- Rex Stout: A simple story about interesting characters is the very best fiction.
- Stephen Wolfram: It is important to be interested in all forms of human knowledge, from the most basic and abstract (such as theoretical physics) to the most concrete and practical (such as starting and running a company).

Author Biography

Samir Varma is a physicist, entrepreneur, and inventor who has always been fascinated by the implications of deterministic physics. He has a Bachelor of Science in Electrical Engineering from Columbia University in 1988, where he was elected to the national engineering honors society, Tau Beta Pi. He earned his Ph.D. in theoretical particle physics from UT Austin in 1993. The renowned physicist E.C.G. Sudarshan was his advisor, and Nobel Laureate, Steven Weinberg, was on his dissertation committee.

Due to the sad cancellation of the Superconducting Super Collider, Samir decided to become an entrepreneur. In 2001, he founded VS Asset Management (www.vsasset.com), a hedge fund manager. After 13 years of research, in 2016, VS Asset created, and is successfully trading, a unique investment methodology called Risk Timing®. He later co-founded Palm Energy Systems (www.palm.energy), which is revolutionizing solar energy technology with aesthetically appealing, functional designs. As Chairman and Chief Scientist at Applango Systems until 2022, Samir pioneered AI algorithms that significantly enhanced efficiency for call centers and telecom companies.

An inventor at heart, Samir holds multiple patents, including innovations in AI, traffic optimization algorithms, automation software, finance, advertising, and medicine. He has also published research in particle physics and chaotic fluid dynamics. Samir is currently working on his second book about economics, finance, and politics. In his spare time, he likes to play squash and guitar, although because of his back, he can no longer play golf. For computationally irreducible reasons, he has a particular fondness for detective and spy stories, and is obsessed with the music of The Beatles and Pink Floyd.

Note to Reader

Thank you for purchasing *The Science of Free Will*. My sincere hope is that you enjoy reading this book as much as I have enjoyed creating it. If you have a few moments, please feel free to add your review of the book at your favorite online site for feedback. Also, if you would like to connect with other books that I have coming in the near future, please visit my website for news on upcoming works, recent blog posts, and to sign up for my newsletter: http://www.samirvarma.com.

Sincerely,
Samir Varma
samir@samirvarma.com

Bibliography

[1] CK-12/Adapted by Christine Miller. *Human Responses to High Altitude*. url: https://humanbiology.pressbooks.tru.ca/chapter/8-8-human-responses-to-high-altitude/.

[2] *35 U.S. Code §101 – Inventions patentable*. July 1952. url: https://www.law.cornell.edu/uscode/text/35/101.

[3] *35 U.S. Code §102 - Conditions for patentability; novelty*. July 1952. url: https://www.law.cornell.edu/uscode/text/35/102.

[4] *35 U.S. Code §103 - Conditions for patentability; non-obvious subject matter*. July 1952. url: https://www.law.cornell.edu/uscode/text/35/103.

[5] *35 U.S. Code §112 - Specification*. July 1952. url: https://www.law.cornell.edu/uscode/text/35/112.

[6] Utku U. Acikalin et al. "Intellectual Property Protection Lost And Competition: An examination using machine learning." November 2022. url: https://www.nber.org/system/files/working_papers/w30671/w30671.pdf.

[7] Anjana Ahuja. "A Machiavellian machine raises ethical questions about AI." In: *Financial Times* (November 2022). url: https://www.ft.com/content/4be7654e-e455-4ce1-8ce6-13736295ef23.

[8] "AI-driven justice may be better than none at all." In: *Financial Times* (2022). url: https://www.ft.com/content/a5709548-03bd-4f65-b9b5-7aa0325c0f6b.

[9] George A. Akerlof. "The Market for 'Lemons': Quality Uncertainty and the Market Mechanism." In: *Quarterly Journal of Economics* 84 (3 1970), pp. 488–500. doi: doi:10.2307/1879431.

[10] Roger MacBride Allen. *Isaac Asimov's Utopia*. Byron Preiss Inc./Ace Books, New York, 1996. ISBN: 0-441-00245-5.

[11] American Physical Society. "Physics and Math Predict Supreme Court Votes." 2022. url: https://web.archive.org/web/20220701084703/https://www.physicscentral.com/explore/plus/supreme-court.cfm.

[12] "Are ants an answer to cancer?" In: *Investor's Business Daily* 39 (44) (6 February 2023), A2.

[13] Philip Ball. "Why free will is beyond physics." 2021. url: https://physicsworld.com/a/why-free-will-is-beyond-physics/.

[14] Gerald R. Baron. "Bottom-Up Physicalism's Impact and Rising Questions." 2020. url: https://medium.com/top-down-or-bottom-up/bottom-up-physicalisms-impact-and-rising-questions-b78016afeae3.

[15] Yoram Bauman. *Mankiw's Ten Principles of Economics, Translated.* 2007. url: https://www.youtube.com/watch?v=VVp8UGjECt4.

[16] Silas R. Beane, Zohreh Davoudi, and Martin J. Savage. "Constraints on the Universe as a Numerical Simulation." 2012. url: https://arxiv.org/pdf/1210.1847.pdf.

[17] Francesco Berto and Jacopo Tagliabue. In: *Stanford Encyclopedia of Philosophy* (2017). url: https://plato.stanford.edu/entries/cellular-automata/#CAFreeWill.

[18] Yudhijit Bhattacharjee. "What are they thinking?" 2022. url: https://apple.news/A-IIm2FbNTym7S7DajemVGw.

[19] Stefan Bode et al. "Tracking the Unconscious Generation of Free Decisions Using Ultra-High Field fMRI". In: *PLoS One* 6.6 (2011), e21612. doi: 10.1371/journal.pone.0021612. url: https://www.ncbi.nlm.nih.gov/pmc/articles/PMC3124546/.

[20] David Brooks. "Do we really know why we do what we do, or is self-awareness a mirage?" In: *Economic Times* (September 2021). url: https://economictimes.indiatimes.com/magazines/panache/do-we-really-know-why-we-do-what-we-do-or-is-self-awareness-a-mirage/articleshow/86317602.cms.

[21] Elizabeth Anne Brown. "Do spiders dream? A new study suggests they do." In: *National Geographic* (2022). url: https://www.nationalgeographic.com/animals/article/jumping-spiders-dream-rem-sleep-study-suggests.

[22] "Bumblebees like ball games." In: *The Economist* (October 2022). url: https://www.economist.com/science-andtechnology/2022/10/27/bumblebees-like-ball-games.

[23] Stephen Bush. "Beware the use of the black box algorithm." In: *Financial Times* (2022), p. 18. url: https://www.ft.com/content/3d5556c5-520e-497a-aa5e-2546c5bc50cf.

[24] D.J. Carragher and P.J.B. Hancock. "Simulated Automated Facial Recognition Systems as Decision-Aids in Forensic Face Matching Tasks." In: *Journal of Experimental Psychology: General* 152.5 (2023), pp. 1286–1304. doi: 10.1037/xge0001310. url: https://psycnet.apa.org/record/2023-24366-001.

[25] Robert Carroll and N. Gregory Mankiw. "The Growth Effects of Tax Policy." 2006. url: http://gregmankiw.blogspot.com/2006/07/growth-effects-of-tax-policy.html.

[26] Sean Carroll. *From Eternity to Here*. Dutton Adult, 2010.

[27] Christopher Chabris and Daniel Simons. *The Invisible Gorilla*. 2010. url: http://www.theinvisiblegorilla.com/gorilla_experiment.html.

[28] Kendra Cherry. "Phineas Gage: His Accident and Impact on Psychology." 2022. url: https://www.verywellmind.com/phineas-gage-2795244.

[29] Adrian Cho. "More evidence to support quantum theory's 'spooky action at a distance.'" August 2015. url: https://www.science.org/content/article/more-evidence-support-quantum-theory-s-spooky-action-distance.

[30] Luis Pedro Coelho. "I am a regional thinker: a review of 'stubborn attachments.'" December 2018. url: https://metarabbit.wordpress.com/tag/tyler-cowen/.

[31] Piers Coleman. "Obituary: Philip W. Anderson (1923–2020)." 2020. url: https://www.nature.com/articles/d41586-020-01318-4.

[32] Conversations with Tyler. "John Cochrane on Economic Puzzles and Habits of Mind (Ep. 117)." 2021. url: https://conversationswithtyler.com/episodes/john-cochrane/.

[33] John Conway and Simon Kochen. "The Free Will Theorem." 2006. url: https://arxiv.org/pdf/quant-ph/0604079.pdf.

[34] Matthew Cook. "Universality in Elementary Cellular Automata." In: *Complex Systems* 15(1) (2004), pp. 1–40. url: https://wpmedia.wolfram.com/uploads/sites/13/2018/02/15-1-1.pdf.

[35] Clive Cookson. "Scientists revive cells and tissues in dead pigs." In: *Financial Times* (2022). url: https://www.ft.com/content/406330fc-6443-4e0f-9022-d9b9567bf45a.

[36] Mo Costandi. "Is the body key to understanding consciousness?" October 2022. url: https://www.theguardian.com/science/2022/oct/02/is-the-body-key-to-understanding-consciousness.

[37] Tyler Cowen. "My new favorite question to ask over lunch." June 2013. url: https://marginalrevolution.com/marginalrevolution/2013/06/my-new-favorite-question-to-ask-over-lunch.html.

[38] Tyler Cowen. *Stubborn Attachments: A Vision for a Society of Free, Prosperous, and Responsible Individuals.* Stripe Press, October 2018.

[39] Tyler Cowen. "The canine model of AGI." February 2023. url: https://marginalrevolution.com/marginalrevolution/2023/02/the-canine-model-of-agi.html.

[40] Ben Crair. "The bizarre bird that's breaking the tree of life." In: *New Yorker* (2022). url: https://www.newyorker.com/science/elements/the-bizarre-bird-thats-breaking-the-tree-of-life.

[41] "Crows Outsmart Monkeys." In: *Investor's Business Daily* 39 (42) (23 January 2023), A2.

[42] DeepMind Technologies. "AlphaFold: a solution to a 50-year-old grand challenge in biology." 2020. url: https://www.deepmind.com/blog/alphafold-a-solution-to-a-50-year-old-grand-challenge-in-biology.

[43] David Deutsch. *The Beginning of Infinity: Explanations That Transform the World*. Viking, 2011.

[44] Keith Devlin. *The Smallest Computer in the World*. November 2007. url: https://web.archive.org/web/20230330054357/https://www.maa.org/external_archive/devlin/devlin_11_07.html.

[45] Dictionary.com. "process." 2022. url: https://www.dictionary.com/browse/process.

[46] Michaeleen Doucleff. "Can dogs smell time? Just ask Donut the dog." December 2022. url: https://www.npr.org/sections/goatsandsoda/2022/12/22/1139781319/can-dogs-smell-time-just-ask-donut-the-dog.

[47] Tim Dowling. "Order force: the old grammar rule we all obey without realising." In: *Guardian* (September 2016). url: https://www.theguardian.com/commentisfree/2016/sep/13/sentence-order-adjectives-rule-elements-of-eloquence-dictionary.

[48] Hayley Dunning. "Bee brains as you have never seen them before." 2016. url: https: //www.imperial.ac.uk/news/171050/bee-brains-have-never-seen-them/.

[49] dvgrn. "Geminoid Challenge." 2013. url: https://conwaylife.com/forums/viewtopic.php?f=2&t=1006&p=9917#p9901.

[50] Adrian Dyer, Jair Garcia, and Scarlett Howard. "Bees can learn higher numbers than we thought – if we train them the right way." October 2019. url: https://phys.org/news/2019-10-bees-higher-thought.html.

[51] Michael Egnor. "Is Free Will an Illusion?" 2018. url: https://evolutionnews.org/2018/07/is-free-will-an-illusion/.

[52] Karen Frances Eng. "Cockroaches, DIY Cyborgs & Mind Control: The Neuro-Revolution Is Coming!" 2015. url: https://fellowsblog.ted.com/theneuro-revolution-is-coming-96ad7171beae.

[53] Ozkan Eren and Naci Mocan. "Emotional Judges and Unlucky Juveniles." In: *American Economic Journal: Applied Economics* 10 (3) (2018), pp. 171–205. doi: 10.1257/app.20160390. url: https://www.aeaweb.org/articles?id=10.1257/app.20160390.

[54] Xiaona Fang and Jin Wang. "Nonequilibrium Thermodynamics in Cell Biology: Extending Equilibrium Formalism to Cover Living Systems." In: *Annual Reviews of Biophysics* 49 (2020), pp. 227–246. url: https://pubmed.ncbi.nlm.nih.gov/32375020/.

[55] Donna Ferguson. "'Bees are really highly intelligent': the insect IQ tests causing a buzz among scientists." In: *Guardian* (2022). url: https://www.theguardian.com/environment/2022/jul/16/bees-are-really-highly-intelligent-the-insect-iq-tests-causing-a-buzz-among-scientists.

[56] Marianne Freiberger and Rachel Thomas. "Phantom jams." 2019. url: https://plus.maths.org/content/phantom-jams.

[57] Robert Frost. "The Road Not Taken." 1915. url: https://www.poetryfoundation.org/poems/44272/the-road-not-taken.

[58] Judy George. "Owning a Pet May Protect Cognitive Health." 2022. url: https://www.medpagetoday.com/meetingcoverage/aan/97337.

[59] Katja Grace. "We don't trade with ants." January 2023. url: https://worldspiritsockpuppet.substack.com/p/we-dont-trade-with-ants.

[60] Michael S.A. Graziano. "Without Consciousness, AIs Will Be Sociopaths." In: *The Wall Street Journal* (January 2023). url: https://www.wsj.com/articles/without-consciousness-ais-will-be-sociopaths-11673619880 (Accessed on 11/01/2023).

[61] Brian Greene. "Quantum mechanics-as currently understood-is deterministic." 2013. url: https://twitter.com/bgreene/status/353879571629740032.

[62] Veronique Greenwood. "Brainless Jellyfish Demonstrate Learning Ability." In: *The New York Times* (September 2023). url: https://www.nytimes.com/2023/09/22/science/jellyfish-learning-neurons.html (Accessed on 11/01/2023).

[63] Brittney Grimes. "A New AI Testing System Could Help Unlock Secrets of the Human Genome." *Interesting Engineering*. Published: December 06, 2022 12:12 PM EST. December 2022. url: https://interestingengineering.com/innovation/ai-testing-system-unlock-secrets-human-genome? (Accessed on 11/01/2023).

[64] Michael Hallermayer. "Researchers answer fundamental question of quantum physics." September 2022. url: https://phys.org/news/2022-09-fundamental-quantum-physics.html.

[65] Jessica Hamzelou. "Neuroscientists have created a mood decoder that can measure depression." In: *MIT Technology Review* (December 2022). url: https://archive.ph/UPflZ.

[66] Karen Hao. "AI is sending people to jail—and getting it wrong." In: *MIT Technology Review* (January 2019). url: https://www.technologyreview.com/2019/01/21/137783/algorithmscriminal-justice-ai/.

[67] Ed Hardy. "M2 MacBook Air runs Windows 11 faster than pricier Dell laptop." August 2022. url: https://www.cultofmac.com/788405/m2-macbook-air-runs-windows-11-faster-than-pricier-dell-laptop/.

[68] Tom Hayden. "The 10 greatest controversies of Winston Churchill's career." 2015. url: https://www.bbc.com/news/magazine-29701767.

[69] Daniela Hernandez. "Raccoons Get a Reputation Makeover: Researchers once considered the critters too rowdy to study. Technology is changing that: 'Little

Einstein raccoons.'" In: *The Wall Street Journal* (December 2022). Photographs by Alana Paterson. url: https://www.wsj.com/articles/raccoons-reputation-makeover-scientists-11670165274 (Accessed on 11/01/2023).

[70] Sabine Hossenfelder. "Einstein on the discreteness of spacetime." *Backreaction*. October 2010. url: https://backreaction.blogspot.com/2010/10/einstein-on-discreteness-of-spacetime.html (Accessed on 11/01/2023).

[71] Scarlett Howard et al. "Honeybees join humans as the only known animals that can tell the difference between odd and even numbers." April 2022. url: https://phys.org/news/2022-04-honeybees-humans-animals-difference-odd.html.

[72] Kenna Hughes-Castleberry. "For the first time, research reveals crows use statistical logic." *Ars Technica*. September 2023. url: https://arstechnica.com/science/2023/09/for-the-first-time-research-reveals-crows-use-statistical-logic/#p3 (Accessed on 11/01/2023).

[73] "India to get more than 100 cheetahs from S. Africa." January 2023. url: https://phys.org/news/2023-01-india-cheetahs-safrica.html.

[74] John S. Bell, in a BBC interview, quoted on Wikipedia. "Superdeterminism." 2022.

[75] Brett J. Kagan et al. "*In vitro* neurons learn and exhibit sentience when embodied in a simulated game-world." In: *Neuron* 110 (October 2022), pp. 1–18. doi: https://doi.org/10.1016/j.neuron.2022.09.001. url: https://www.cell.com/neuron/fulltext/S0896-6273(22)00806-6.

[76] Julia Kagan. "Lump of Labor Fallacy." In: *Investopedia* (September 2022). url: https://www.investopedia.com/terms/l/lump-of-labour-fallacy.asp.

[77] Natalie L. Kahn and Jing-Jing Shen. "Brian Greene Talks the Physics of Free Will at Science Center Lecture." 2020.

url: https://www.thecrimson.com/article/2020/2/20/brian-greene-physics-will/.

[78] Claudia Kallmeier. "New fur for the quantum cat: Entanglement of many atoms discovered for the first time." September 2022. url: https://phys.org/news/2022-09-fur-quantum-cat-entanglement-atoms.html.

[79] Holden Karnofsky. "Why Would AI 'Aim' To Defeat Humanity?" November 2022. url: https://www.cold-takes.com/why-would-ai-aim-to-defeat-humanity/.

[80] John Kay. "Obliquity." 2004. url: https://www.johnkay.com/2004/01/17/obliquity/.

[81] James Kerr. "How Bill Belichick's 'Do Your Job' Mantra Applies to Leadership." 2015. url: https://www.inc.com/james-kerr/how-do-your-job-can-be-a-difference-maker-for-your-company.html.

[82] Evan Kozliner. "Algorithmic Beauty: An Introduction to Cellular Automata." 2019. url: https://bit.ly/2pvYjAE.

[83] Ville Kuosmanen. "I Played Chess Against ChatGPT." *Medium*. 2023. url: https://villekuosmanen.medium.com/i-played-chess-against-chatgpt-4c2cc78b5acf (Accessed on 11/01/2023).

[84] Esther Landhuis. "Is Death Real? A mind-blowing scientific discovery could change what it means to die." In: *Popular Mechanics* (December 2022). url: https://apple.news/AMuw9rRj1RnGYzFxi-X_SPg.

[85] Rolf Landua and Marlene Rau. "The LHC: a step closer to the Big Bang." In: *Science in School: The European Journal for science teachers* 10 (2008). url: https://www.scienceinschool.org/article/2008/lhcwhy/.

[86] Nick Lane. *Transformer: The Deep Chemistry of Life and Death.* W.W. Norton & Company, May 2022. Kindle Edition.

[87] LifeWiki. "Gemini." 2022. url: https://conwaylife.com/wiki/Gemini.

[88] Seth Lloyd. "A theory of quantum gravity based on quantum computation." 2018. url: https://arxiv.org/pdf/quant-ph/0501135.pdf.

[89] Roger Lowenstein. "Fighting Inequality With a Minimum of 'Leaks.'" In: *The Wall Street Journal* (2022). url: https://www.wsj.com/articles/fightingin-equality-with-a-minimum-of-leaks-11659708045.

[90] N. Gregory Mankiw. "Remarks by Dr. N. Gregory Mankiw, Chairman Council of Economic Advisers, at the National Bureau of Economic Research Tax Policy and the Economy Meeting" (at the National Press Club). 2003. url: https://georgewbush-whitehouse.archives.gov/cea/NPressClub20031104.html.

[91] N. Gregory Mankiw. "Remarks of Dr. N. Gregory Mankiw, Chairman Council of Economic Advisers, at the Annual Meeting of the National Association of Business Economists Atlanta, Georgia." 2003. url: https://georgewbushwhitehouse.archives.gov/cea/mankiw_speech_nabe_20030915.html.

[92] N. Gregory Mankiw. "Yes, r > g. So What?" 2015. url: https://scholar.harvard.edu/files/mankiw/files/yes_r_g_so_what.pdf.

[93] Rick Marshall. "Obscenity Case Files: Jacobellis v. Ohio ('I know it when I see it')." url: https://cbldf.org/about-us/case-files/obscenity-case-files/obscenity-case-files-jacobellis-v-ohio-i-know-it-when-i-see-it/.

[94] Oleg Mazonka and Alex Kolodin. "A Simple Multi-Processor Computer Based on Subleq." 2011. url: https://arxiv.org/pdf/1106.2593.pdf.

[95] Jonathan McDowell. "The public are excited by the images, but astronomers are as excited by the spectra[...]." 2022. url: https://twitter.com/planet4589/status/1546973445510119424.

[96] Michael McKenna and D. Justin Coates. "Compatibilism." 2019. url: https://plato.stanford.edu/entries/compatibilism/.

[97] Chris Melore. "Are you smarter than AI? Computer language model the clear winner over people in IQ tests." December 2022. url: https://studyfinds.org/are-you-smarter-than-ai/.

[98] El-Mahdi El-Mhamdi et al. "On the Impossible Safety of Large AI Models." Version v2. September 2022. doi: 10.48550/arXiv.2209.15259. arXiv: 2209.15259[cs.LG]. url: https://doi.org/10.48550/arXiv.2209.15259 (Accessed on 11/01/2023).

[99] Aaron Mok. "Getting Emotional with ChatGPT Could Get You the Best Outputs." In: *Business Insider* (November 2023). url: https://www.businessinsider.com/chatgpt-llm-ai-responds-better-emotional-language-prompts-study-finds-2023-11 (Accessed 11/07/2023).

[100] More Famous Quotes. "Quantum Chemistry Quotes." 2022. url: https://www.morefamousquotes.com/topics/quantum-chemistry-quotes/.

[101] Dominique Mosbergen. "Crows Found to Be Smarter Than We Think: Two birds learned to organize structures in a way once thought unique to humans, researchers said." In: *The Wall Street Journal* (November 2022). url: https://www.wsj.com/articles/crows-found-to-be-smarter-than-we-think11667412092.

[102] Richard Muller. "If the universe follows causality, how can there be free will?" url: https://www.quora.com/If-the-universe-follows-causality-how-can-there-be-free-will/answer/Richard-Muller-3.

[103] Richard Muller. *Now: The Physics of Time.* W.W. Norton & Company, 2016.

[104] Michael T. Murphy et al. "A limit on variations in the fine-structure constant from spectra of nearby Sun-like stars."

In: *Science* 378.6620 (November 2022). url: https://www.science.org/doi/10.1126/science.abi9232.

[105] Steve Nadis. "This is your brain. This is your brain on code." December 2022. url: https://news.mit.edu/2022/your-brain-your-brain-code-1221.

[106] NASA. "NASA's Webb Detects Carbon Dioxide In Exoplanet Atmosphere." August 2022. url: https://www.nasa.gov/feature/goddard/2022/nasa-s-webb-detects-carbon-dioxide-in-exoplanet-atmosphere.

[107] National Geographic. "Animals dream too—here's what we know." 2022. url: https://www.nationalgeographic.com/animals/article/animals-dream-too-heres-what-we-know.

[108] NOAA. "How Reliable Are Weather Forecasts?" 2022. url: https://web.archive.org/web/20220629153841/scijinks.gov/forecastreliability/.

[109] Rachel Nuwer. "This Japanese Shrine Has Been Torn Down And Rebuilt Every 20 Years for the Past Millennium." October 2013. url: https://www.smithsonianmag.com/smart-news/this-japanese-shrine-has-been-torn-down-and-rebuilt-every-20-years-for-the-past-millennium-575558/.

[110] Meghan O'Gieblyn. "Can Robots Evolve Into Machines of Loving Grace?" In: *Wired* (2021). url: https://www.wired.com/story/can-robots-evolve-into-machines-of-loving-grace/.

[111] A.W. Ohlheiser. "How to befriend a crow." In: *MIT Technology Review* (October 2022). url: https://www.technologyreview.com/2022/10/31/1062370/how-to-befriend-a-crow-crowtok-tiktok.

[112] Gustavus A. Ohlinger. "WSC: A midnight interview, 1902." 2015. url: https://winstonchurchill.org/publications/finest-hour/finest-hour159/wsc-a-midnight-interview-1902/.

[113] Arthur M. Okun. *Equality and Efficiency: The Big Tradeoff (A Brookings Classic)*. Brookings Institution Press; Revised edition (April 30, 2015), 2015.

[114] Chris Opfer. "Does your body really replace itself every seven years?" 2021. url: https://science.howstuffworks. com/life/cellular-microscopic/does-body-really-replace-seven-years.htm.

[115] Darren Orf. "It Took 12 Years To Completely Map a Baby Fruit Fly's Brain." In: *Popular Mechanics* (March 2023). url: https://www.popularmechanics.com/science/animals/a43266443/fruit-fly-brain/?utm_source=tldrnewsletter.

[116] Michael Peel. "Smart rats show human-like powers of imagination in neural research (online); Rats credited with brains that possess human-like imagination (print)." In: *Financial Times* (November 2023). url: https://www.ft.com/content/5374e443-6523-49be-bc55-61febb9a92e9 (Accessed 11/01/2023).

[117] "Pigs reconcile after fighting. That confirms their intelligence: Bystanders offer consolation to losers." In: *The Economist* (November 2022). url: https://www.economist.com/science-and-technology/2022/11/09/pigs-reconcile-after-fighting-that-confirms-their-intelligence.

[118] Cell Press. "Dogs cry more when reunited with their owners." 2022. url: https://www.eurekalert.org/news-releases/962142.

[119] Association for Psychological Science. "People Neglect Who They Really Are When Predicting Their Own Future Happiness." 2011. url: https://www.psychologicalscience.org/news/releases/people-neglect-who-they-really-are-when-predicting-their-own-future-happiness.html.

[120] Sofia Quaglia. "Crows can understand probability like primates do." In: *New Scientist* (July 2023). url: https://www.newscientist.com/article/2381335-crows-can-understand-probability-like-primates-do/ (Accessed on 11/01/2023).

[121] Elena Renken. "A New Doorway to the Brain." October 2022. url: https://nautil.us/a-new-doorway-to-the-brain-242099/.

[122] Raúl Rojas. "A Tutorial Introduction to Lambda Calculus."
1998. url: https://personal.utdallas.edu/~gupta/courses/
apl/lambda.pdf.

[123] Hanno Sauer. "The end of history." In: *Inquiry* (2022). doi:
doi:10.1080/0020174X.2022.2124542. url: https://www.
tandfonline.com/doi/full/10.1080/0020174X.2022.2124542.

[124] William Shakespeare. *Macbeth*. Act 5, Scene 5. 1606.

[125] Amit Shraga. "The Body's Elements." 2020. url: https://
davidson.weizmann.ac.il/en/online/orderoutofchaos/
body%E2%80%99selements#home.

[126] Jascha Sohl-Dickstein. "The Hot Mess Theory of AI
Misalignment: More Intelligent Agents Behave Less
Coherently." Jascha Sohl-Dickstein's Blog. March 2023.
url: https://sohl-dickstein.github.io/2023/03/09/coherence.
html (Accessed on 11/01/2023).

[127] Jessica Stillman. "How Amazon's Jeff Bezos Made One of
the Toughest Decisions of His Career." 2016. url: https://
www.inc.com/jessica-stillman/jeff-bezos-this-is-how-to-
avoid-regret.html.

[128] Kevin Stroud. *History of English Podcast (Ep. 117)*. 2021. url:
https://historyofenglishpodcast.com/wp-content/uploads/
2021/10/HOETranscript-Episode058.pdf.

[129] Alex Tabarrok. "AGI is Coming." *Marginal Revolution*. March
2023. url: https://marginalrevolution.com/marginalrevolution/
2023/03/agi-is-coming.html?utm_source=feedly&utm_
medium=rss&utm_campaign=agi-is-coming (Accessed on
11/01/2023).

[130] Alex Tabarrok. "Humans Will Align with the AIs
Long Before the AIs Align with Humans." *Marginal
Revolution*. February 2023. url: https://marginalrevolution.
com/marginalrevolution/2023/02/aiporn.html?utm_
source=feedly&utm_medium=rss&utm_campaign=ai-
porn (Accessed on 11/01/2023).

[131] Alex Tabarrok. "The End of History (of Philosophy)." 2022. url: https://marginalrevolution.com/marginalrevolution/2022/09/the-end-of-history-of-philosophy.html.

[132] Max Tegmark. "The Multiverse Hierarchy." 2005. url: https://arxiv.org/pdf/0905.1283.pdf.

[133] Gillian Tett. "AI will not remedy all our real estate woes." In: *Financial Times* (March 2023). url: https://www.ft.com/content/80306709-f26d-4d8f-85b7-652ec17cb7c4 (Accessed on 11/01/2023).

[134] The Physics arXiv Blog. "AI Chatbot Spontaneously Develops A Theory of Mind." *Discover Blogs*. February 2023. url: https://www.discovermagazine.com/mind/ai-chatbot-spontaneously-develops-a-theory-of-mind (Accessed on 11/01/2023).

[135] John Tilson. "2018 Stupid Wine Description Winners." In: *The Underground WineLetter* (January 2019). url: https://www.undergroundwineletter.com/2019/01/2018-stupid-wine-description-winners/.

[136] Harvard University. "How squid and octopus get their big brains." November 2022. url: https://phys.org/news/2022-11-squid-octopus-big-brains.html.

[137] *US Constitution, Article I Section 8 Clause 8.* url: https://fairuse.stanford.edu/law/us-constitution/.

[138] Steven Weinberg. *The First Three Minutes: A Modern View of the Origin of the Universe.* Basic Books, 1977.

[139] David Weinberger. "AI's ways of being immoral." July 2022. url: https://www.kmworld.com/Articles/Columns/Perspective-onKnowledge/AIs-ways-of-being-immoral-153629.aspx.

[140] Chloe Weise, Christian Cely Ortiz, and Elizabeth A. Tibbetts. "Paper wasps form abstract concept of 'same and different.'" In: *Proceedings of the Royal Society B* (July 2022). doi: 10.1098/rspb.2022.1156.

[141] Wikipedia. "A New Kind of Science: Principle of Computational Equivalence." 2022. url: https://en.wikipedia.org/wiki/A_New_Kind_of_Science#Principle_of_computational_equivalence.

[142] Wikipedia. "Anomalous magnetic dipole moment." 2022. url: https://en.wikipedia.org/wiki/Anomalous_magnetic_dipole_moment.

[143] Wikipedia. "Computational Irreducibility." 2021. url: https://en.wikipedia.org/wiki/Computational_irreducibility.

[144] Wikipedia. "Condorcet Paradox." 2022. url: https://en.wikipedia.org/wiki/Condorcet_paradox.

[145] Wikipedia. "Demonstration of Sensitivity to Initial Conditions in a Double Pendulum." 2022. url: https://en.wikipedia.org/wiki/Double_pendulum.

[146] Wikipedia. "Epimenides paradox." 2022. url: https://en.wikipedia.org/wiki/Epimenides_paradox.

[147] Wikipedia. "Fallacy of Composition." 2022. url: https://en.wikipedia.org/wiki/Fallacy_of_composition.

[148] Wikipedia. "Isomorphism." 2022. url: https://en.wikipedia.org/wiki/Isomorphism.

[149] Wikipedia. "Markman Hearing." 2021. url: https://en.wikipedia.org/wiki/Markman_hearing.

[150] Wikipedia. "Newcomb's paradox." 2021. url: https://en.wikipedia.org/wiki/Newcomb%27s_paradox.

[151] Wikipedia. "NeXTStep." 2022. url: https://en.wikipedia.org/wiki/NeXTSTEP.

[152] Wikipedia. "Precision Tests of QED." 2022. url: https://en.wikipedia.org/wiki/Precision_tests_of_QED.

[153] Wikipedia. "Rule 110 cellular automaton." 2021. url: https://en.wikipedia.org/wiki/Rule_110.

[154] Wikipedia. "Rule 30 cellular automaton." 2021. url: https://en.wikipedia.org/wiki/Rule_30.

[155] Wikipedia. "Three Laws of Robotics." 2022. url: https://en.wikipedia.org/wiki/Three_Laws_of_Robotics.

[156] Wikipedia. "Trolley problem." 2022. url: https://en.wikipedia.org/wiki/Trolley_problem.

[157] Wikipedia. "Turritopsis dohrnii." 2022. url: https://en.wikipedia.org/wiki/Turritopsis_dohrnii.

[158] Wikiversity. "Introduction to Turing Machines." 2022. url: https://en.wikiversity.org/wiki/Introduction_to_Turing_Machines.

[159] Frank Wilczek. "The Power and Poetry of Random Digits." In: *The Wall Street Journal* (June 2021), p. C4. url: https://on.wsj.com/3iCIr9m.

[160] Alex Wilkins. "Wasps can grasp abstract concepts such as 'same' and 'different.'" In: *New Scientist* (July 2022). url: https://www.newscientist.com/article/2329327-wasps-can-grasp-abstract-concepts-such-as-same-and-different/.

[161] Stephen Wolfram. *A New Kind of Science*. Wolfram Media, 2002. url: https://www.wolframscience.com/nks/.

[162] Stephen Wolfram. "Combinators and the Story of Computation." 2020. url: https://writings.stephenwolfram.com/2020/12/combinators-and-the-story-of-computation/.

[163] Stephen Wolfram. "The Concept of the Ruliad." 2021. url: https://writings.stephenwolfram.com/2021/11/the-concept-of-the-ruliad/.

[164] Stephen Wolfram. "The Wolfram 2,3 Turing Machine Research Prize, Technical Details." url: https://www.wolframscience.com/prizes/tm23/technicaldetails.html.

[165] Wolfram Research. "Tech Note: Random Number Generation." 2021. url: https://reference.wolfram.com/language/tutorial/RandomNumberGeneration.html#185956823.

[166] Charlie Wood. "Physicists Rewrite a Quantum Rule That Clashes With Our Universe." In: *Quanta Magazine* (September 2022). url: https://www.quantamagazine.org/physicists-rewrite-a-quantum-rule-that-clashes-with-our-universe-20220926/.

[167] Charlie Wood. "Powerful 'Machine Scientists' Distill the Laws of Physics From Raw Data." In: *Quanta Magazine* (May 2022). url: https://www.quantamagazine.org/machine-scientists-distill-the-laws-of-physics-from-raw-data-20220510/.

[168] Aylin Woodward. "Electrical Brain Stimulation Improves Memory, New Study Shows." In: *The Wall Street Journal* (2022). url: https://www.wsj.com/articles/electrical-brain-stimulation-improves-memory-new-study-shows-11661180400.

[169] Bob Yirka. "Honeybees show withdrawal symptoms when weaned off alcohol." June 2021. url: https://phys.org/news/2021-06-honeybees-symptoms-weaned-alcohol.html.

[170] Bob Yirka. "Honeybees use a 'mental number line' to keep track of things." October 2022. url: https://phys.org/news/2022-10-honeybees-mental-line-track.html.

IFF
BOOKS

ACADEMIC AND SPECIALIST

Iff Books publishes non-fiction. It aims to work with authors and titles that augment our understanding of the human condition, society and civilisation, and the world or universe in which we live. If you have enjoyed this book, why not tell other readers by posting a review on your preferred book site.
Recent bestsellers from Iff Books are:

Why Materialism Is Baloney
How true skeptics know there is no death and fathom answers to life, the universe, and everything
Bernardo Kastrup
A hard-nosed, logical, and skeptic non-materialist metaphysics, according to which the body is in mind, not mind in the body.
Paperback: 978-1-78279-362-5 ebook: 978-1-78279-361-8

The Fall
Steve Taylor
The Fall discusses human achievement versus the issues of war, patriarchy and social inequality.
Paperback: 978-1-78535-804-3 ebook: 978-1-78535-805-0

Brief Peeks Beyond
Critical essays on metaphysics, neuroscience, free will, skepticism and culture
Bernardo Kastrup
An incisive, original, compelling alternative to current mainstream cultural views and assumptions.
Paperback: 978-1-78535-018-4 ebook: 978-1-78535-019-1

Framespotting

Changing how you look at things changes how
you see them
Laurence & Alison Matthews
A punchy, upbeat guide to framespotting. Spot deceptions and
hidden assumptions; swap growth for growing up. See and be free.
Paperback: 978-1-78279-689-3 ebook: 978-1-78279-822-4

Is There an Afterlife?

David Fontana
Is there an Afterlife? If so what is it like? How do Western ideas
of the afterlife compare with Eastern? David Fontana presents the
historical and contemporary evidence for survival of
physical death.
Paperback: 978-1-90381-690-5

Nothing Matters

a book about nothing
Ronald Green
Thinking about Nothing opens the world to everything by
illuminating new angles to old problems and stimulating new
ways of thinking.
Paperback: 978-1-84694-707-0 ebook: 978-1-78099-016-3

Panpsychism

The Philosophy of the Sensuous Cosmos
Peter Ells
Are free will and mind chimeras? This book, anti-materialistic but
respecting science, answers: No! Mind is foundational
to all existence.
Paperback: 978-1-84694-505-2 ebook: 978-1-78099-018-7

Punk Science
Inside the Mind of God
Manjir Samanta-Laughton
Many have experienced unexplainable phenomena; God, psychic
abilities, extraordinary healing and angelic encounters. Can
cutting-edge science actually explain phenomena
previously thought of as 'paranormal'?
Paperback: 978-1-90504-793-2

The Vagabond Spirit of Poetry
Edward Clarke
Spend time with the wisest poets of the modern age and of the
past, and let Edward Clarke remind you of the importance of
poetry in our industrialized world.
Paperback: 978-1-78279-370-0 ebook: 978-1-78279-369-4

Readers of ebooks can buy or view any of these bestsellers by
clicking on the live link in the title. Most titles are published in
paperback and as an ebook. Paperbacks are available in traditional
bookshops. Both print and ebook formats are available online.
Find more titles and sign up to our readers' newsletter at
www.collectiveinkbooks.com/non-fiction
Follow us on Facebook at
www.facebook.com/CINonFiction

Printed and bound by CPI Group (UK) Ltd, Croydon, CR0 4YY

13/01/2025

01819353-0001